individualized fitness programs

46869

Frank Vitale

Associate Supervisor of Physical Education
University of California, San Diego

Prentice-Hall, Inc.
Englewood Cliffs, New Jersey

Library of Congress Cataloging in Publication Data

VITALE, FRANK.
   Individualized fitness programs.

   Includes bibliography
     1. Exercise—Physiological effect. 2. Posture.
3. Physical fitness. I. Title.
QP301.V58   613.7   72–7406
ISBN 0–13–457002–2
ISBN 0–13–456996–2 (pbk)

to jan and our children—
gina, dana, kara, and frankie

© 1973 by Prentice-Hall, Inc.
Englewood Cliffs, New Jersey

10 9 8 7 6 5 4 3 2 1

Printed in the United States of America

Prentice-Hall International, Inc. *London*
Prentice-Hall of Australia, Pty. Ltd., *Sydney*
Prentice-Hall of Canada, Ltd., *Toronto*
Prentice-Hall of India Private Limited, *New Delhi*
Prentice-Hall of Japan, Inc. *Tokyo*

contents

3
effects and benefits of exercise   52

4
exercise principles, techniques, and precautions   73

5
individual exercise programs   107

foreword

Much has been said and written about the role of exercise in promoting and maintaining good health. Research in medicine, physiology, and related disciplines eliminates any doubt that sound programs for personal physical fitness are necessary, yet present-day thinking on physical fitness still leaves much to be desired. To a great extent, many programs today lack the balance that comes from viewing all the aspects of an individual's well-being: physical, intellectual, social, and spiritual.

There are two extreme opinions regarding the role of physical fitness in a person's life. Some consider the individual a wholly intellectual being; others see him or her as only a body. In reality, each person is a unique blend of intellectual, physical, emotional, and social characteristics interacting to form a complete human being. Failure to recognize the interaction and unity of all aspects or over-emphasis or neglect of any one is a failure to consider the whole person. Throughout this book, the author has attempted to regard physical fitness from an interacting and unified point of view, so that all aspects harmonize.

People in all walks of life are becoming more and more cognizant of the role of exercise, sport and recreational activities, diet, and other important health practices in affecting their total well-being. There is greater concern than ever before for the growing need to develop habits that will lead to a lifetime of good health and personal happiness. This book makes a valuable contribution toward developing these habits.

The author, Frank Vitale, has devoted a lifetime to the study and practice of total fitness. His students invariably attest to the soundness of his own program at the University of California at San Diego and share the author's concern that his fitness message will reach beyond the U. C. San Diego campus.

TED FORBES
Chairman, Department
of Physical Education,
University of California,
San Diego

preface

The purpose of this book is to provide a complete and practical source of information and programs that will help any person develop and maintain sound, lifetime habits of fitness. It is designed to enable the reader of any age or sex to plan and progressively evaluate a personalized, permanent fitness program based on an understanding of the most current knowledge and principles of physical fitness. As a matter of fitness philosophy, the book includes a focus not only on exercise and fitness, but on other important areas as well, including: posture and body mechanics; diet, nutrition, and weight control; rest, relaxation, and recreation; and sound health habits. The aim is to develop a broad and practical approach to fitness in a way that avoids "preaching fitness," recognizes today's value systems, and leads to an understanding of the many aspects of fitness and its necessity as an integral part of everyone's lifestyle.

Chapter 1 establishes this broader concept of fitness and the need for it on a national and individual level. The major problem of cardiovascular disease and other major health problems are tied in with their identified risk factors. The concept and components of fitness to be developed are then summarized to set the stage for a detailed treatment of how to overcome or minimize these risks. Chapter 2 provides a personal method of continuous evaluation in terms of the multiple facets of fitness presented and includes such items as body build, posture, body proportions and body fat analysis, cardiovascular and respiratory fitness, motor fitness, sports activity competency, ability to relax evaluation, general health and fitness checks, and evaluations of smoking and drinking habits.

Chapters 3 through 6 place specific emphasis on the role of exercise in fitness. Chapter 3 presents the specific effects on and benefits of exercise for the body and includes a brief, nontechnical explanation of the physiological functions of the body's major systems in order to provide a basis for understanding the exercise principles and programs that follow. Chapter 4 gives practical guides for the selection, use, and development of exercise programs with respect to purpose, type, and effectiveness of various exercises, logical progression, and safety precautions. A variety of eight individual exercise programs are described and evaluated in Chapter 5; those that meet specific needs and interests can be selected from this section. These programs include calisthenics, weight training, isometric exercises, iso-kinetic exercises, aerobic exercises, the Royal Canadian Air Force Exercise Plans, the P-F Plan, and water exercises. A further application of specific principles and exercise programs for the correction of postural defects and the alleviation of low back pain is covered in Chapter 6.

Chapter 7 gives the essentials of effective weight control and shows how to develop a sound personal weight control plan utilizing a combination of proper exercise, diet, and nutrition. The recently recognized effects of stress and tension on mental and physical well-being are discussed in Chapter 8, followed by specific psychological and physical relaxation techniques aimed at reducing their harmful effects. Chapter 9 concludes by pointing out the use of various sport and recreational activities in a total program of physical fitness. General guides are given to help select those activities best suited for individual needs and interests, together with specific principles and guides for learning these activities.

Throughout the book an effort is made to develop a deeper understanding of fitness. The emphasis is not only on the "how," but also the "why." Reasons are given in terms of how the body operates, how it responds to various sound and unsound practices, and why certain principles and techniques are necessary. Without such an understanding, the "how" loses its meaning; what is meaningless becomes useless. Technical terms have been deliberately avoided or defined when necessary, and an attempt has been made to present the information and programs in a direct, concise, and easy-to-understand manner so that they can be of maximum practical value.

acknowledgements

Sincere thanks are due to many persons without whose help this book would not have been possible: Dr. Herbert F. Stallworth, former President at College of the Mainland, Texas City, Texas and current Assistant to the President at Florida Atlantic University for the opportunity as Chairman of the Physical Education Department to develop the ideas upon which this book is based; Dr. Fred Taylor, former Dean of Instruction and now President at College of the Mainland; Miss Jane Meredith and Mr. Ron Ummel whose assistance and practical suggestions helped in the initial development of the manuscript. At the University of California, San Diego, I would like to thank Ted Forbes for his encouragement and material contributions to certain sections in the chapters on Principles of Exercise, Posture, and Handling Stress and Tension; Howard Hunt for his research contributions to the sections on Evaluation, Exercise Programs, and Weight Control and other valuable and practical suggestions; Lee Johnson, Bert Kobayashi, and Jim White for reading sections of the manuscript and for their constructive criticisms; and the secretarial staff for their assistance in typing the manuscripts: Miss Lorraine Kimball, Mrs. Shirley Mehas, Mrs. Maureen Wood, Mrs. Victoria Spencer, Mrs. Jane Shipkey, and Mrs. Marj Javet. Finally, to Dr. Paul Governali, Professor of Physical Education at San Diego State College for his editing and comments on the initial manuscript. And I am indebted to two men for the fitness philosophy and approach expressed in this book: Dr. Harry A. Scott and Dr. Clifford L. Brownell, my former professors and advisors at Teacher's College, Columbia University.

FRANK VITALE

1

the need
for a broader concept
of fitness

Fitness has become a national concern. At a time when our nation is enjoying the highest standard of living in the world, health problems are rising to epidemic proportions. Despite advances in technology and medicine, there is ample evidence that the quality of life in America leaves much to be desired. In terms of longevity alone, there has been no significant increase in the life expectancy of the American male in the last twenty years. The men of at least twenty-six nations and the women of eleven countries can look forward to a longer life expectancy at age forty-five than their counterparts in America. In addition to early deaths, temporary and chronic illnesses and disabilities sap the strength and vitality of our nation. Over 28 million Americans have some degree of disability. More than 9 million children under fifteen years of age have one or more chronic conditions. Millions of dollars are lost daily from job absenteeism, and billions are spent in hospital and medical care. Much of this tragic draining of our human and economic resources is the result of factors that require a broader look at physical fitness and methods for achieving it.

Basically, fitness means being in good physical condition and being able to function at one's best level. But more than the body is involved. Total fitness for living necessarily involves spiritual, mental, emotional, and social, as well as physical, qualities. Each is dependent upon and affected by the other. Though our primary concern is developing good physical condition, we must recognize this interrelatedness and inter-dependence in our approach. It must be recognized not only in terms of the complexity of the mutual effects created, but also in its implication that understanding is needed if efforts are to be meaningful and lasting.

We must also recognize that fitness is not a static condition, but a dynamic one that is constantly changing and is influenced by many factors. Basic to good physical condition is good medical and dental care; the proper type, amount, and method of exercise; good posture and body mechanics in daily living; proper diet and weight control; adequate rest, relaxation, and recreation; and sound practices with respect to drinking, smoking, and the use of drugs. Weakness or neglect in any of these areas can have a detrimental effect on physical condition and undermine the effectiveness of efforts in the other areas. Because of the interrelated-ness of physical fitness with all areas of total fitness and the multiple factors involved in it, the term *fitness* must be seen as implying more than just the "physical" and "exercise."

Due to ignorance or neglect in some of these areas, our hospitals and cemeteries today are filled with people who shouldn't be there—at least not yet. We will now take a look at some of the major health problems that have caused concern for a broader concept of fitness and the need for achieving it.

cardiovascular disease

Cardiovascular disease includes a group of diseases relating to the heart and blood vessels or the blood circulatory system. It is manifested in a decreased blood supply to vital organs and parts of the body. When the blood supply to a particular part of the body is cut off, that part is damaged. Death or crippling disability may follow. When blood supply to the heart is cut off, a heart attack is the result. When it happens in the brain, a stroke occurs. If the lung is affected, both the lung and heart are damaged. Restricted blood supply to the brain and legs is often the cause of senility or cripplement in the aged.

The United States has the highest death rate from cardiovascular disease of any nation in the world. According to a 1971 report of the National Education Health Committee, diseases of the heart and blood vessels account for over half (53 percent) of all deaths—or about 1 million annually. The number of persons partially or totally disabled by cardiovascular disease is also staggering, and no age or sex is immune from it. Of the 22 million Americans afflicted with the disease, approximately 500,000 are of school age, and, during the past few years, there has been a significant increase in circulatory disease among young males aged fifteen to forty-five. Among those who have died from cardiovascular disease in recent years, 35 percent were under age sixty-five. Women, protected by certain hormones, are less susceptible to the disease prior to menopause. Between the ages of forty-five and sixty-two, however, the ratio of heart attacks between men and women—one manifestation of cardiovascular disease—drops from 13 to 1 to a 2 to 1 ratio. After sixty-five, the ratio is equal.

atherosclerosis

The most common cardiovascular disease and the underlying cause of most heart attacks and strokes is atherosclerosis. *Atherosclerosis* is a form of *arteriosclerosis* (hardening of the arteries) in which cholesterol and other fatty deposits become attached to the linings of arteries and interfere with blood circulation. Atherosclerosis of the legs, in addition to causing crippling pain, may lead to gangrene and amputation. The roughened surface of the artery walls anywhere in the circulatory system may cause a blood clot to form. If the blood clot, or a part of it which may break away, blocks off an artery, the body part served by that artery loses its blood supply and becomes damaged. When it happens in an artery supplying the heart or brain, a coronary heart attack or stroke results. In medical terms, the forming of a blood clot which results in the blocking of an artery is called *thrombosis*. When a blood clot circulating in the blood becomes lodged in an artery, it is called an *embolism*.

Atherosclerosis, unlike hardening of arteries and loss of elasticity due to *hypertension* (high blood pressure) is not primarily associated with old age. Though it may not become a problem until middle-age, there is evidence that it can begin to develop at an early age and affects the entire body. The first indications of atherosclerotic development in the young came from autopsies performed on American soldiers killed during World War II. Many in their early twenties were found to have the disease. Similar autopsies performed during the Korean War showed the disease to be even more widespread. Seventy-seven percent of the men, whose average age was only twenty-two, had evidence of fatty deposits in their coronary arteries. More recently, fatty deposits have been found in the arteries of teenagers and children.

There is evidence to indicate that atherosclerosis is a product of the more highly developed and affluent nations of the Western world. Autopsies performed on Japanese soldiers during the Korean War showed far less evidence of artery deterioration. More recent studies of peoples in underdeveloped countries of Africa show that fatty deposits in the walls of arteries are almost unknown. In the leading industrial nations of the Western world, however, cardiovascular disease is listed among the ten leading causes of death. These findings have led many medical authorities to suspect that differences in diet, level of physical activity, smoking, and other living habits are major factors in the development of atherosclerosis and its ultimate complications: heart attacks, strokes, and premature disability.

heart disease

Over 500,000 deaths occur from heart attacks each year. There are over twenty-one varieties of heart disease. Two of the commonest forms are *coronary heart disease* and *hypertensive heart disease*. Coronary heart disease results from an obstruction of the blood supply to the heart muscle itself in one or both of its major arteries (coronary arteries). Hypertensive heart disease is associated with high blood pressure, or hypertension, an excessive amount of pressure placed on the arteries as the heart pumps blood through them. Hypertensive heart disease results from some narrowing of the arteries, causing the heart to overwork and ultimately to wear out.

Though there is still much to be learned about both the causes and cures for heart attacks, medical studies have isolated certain conditions or living habits that are associated with heart attacks. The conditions causing strokes and other forms of cardiovascular disease are similar. The most significant study is the Framingham study in which sixteen years of data were collected from 1955 to 1970 from men and women aged thirty to sixty-two. According to Dr. William B. Castelli, former director of laboratories in the Framingham Heart Program, the findings show we are having an epidemic. From the data, certain conditions or living habits associated with the incidence of heart attacks were identified and labeled as "risk factors." The presence of only one factor increases the chance of having a heart attack, and combinations of two or more multiply the risk considerably. The major risk factors and the numerical risk of a heart attack associated with them are:

1. High levels of cholesterol or other fatty deposits in the blood (3–6 times)
2. Overweight (2–3 times)
3. Diabetes (2–4 times)

4. High blood pressure (8 times)
5. Lack of exercise (2 times)
6. Smoking (3–6 times)

Any combination of three of these factors may increase the risk up to 30 times. Other conditions found to be associated with a higher risk of heart attack are: a family history of heart attacks in middle age; a spouse with coronary heart disease; degree of alcohol consumption; and low vital capacity (lung capacity).

High cholesterol and other fatty substances. High levels of cholesterol and other fatty deposits in the blood are associated with the development of atherosclerosis and the risk of heart attack, stroke, and other cardiovascular disease. Cholesterol is not a foreign substance, but a very important constituent of the body. It is reflected in some of the most complex hormones and vitamins that are essential for life and health.

Diet is the most important single factor in the production of cholesterol—the easiest to detect and measure—and other fatty acids. In the United States and other countries with a high standard of living, some of the most widely used foods are rich in saturated fats, cholesterol, calories, and concentrated carbohydrates. These foods tend to raise the level of both cholesterol and other fatty deposits in the blood. They include certain meats (beef, pork, lamb), dairy products (whole milk, butter, eggs), and foods high in refined sugars (syrups, jellies, sugared drinks). The body also synthesizes or creates cholesterol from simpler dietary components in response to its needs. Thus, even if the diet contained no cholesterol, the body would still produce it.

The main concern is how much cholesterol is in the blood. Controversy still exists over what constitutes a "normal" cholesterol level. Furthermore, because of other factors involved in its production and use, it is considered almost impossible to obtain a stable reading or to completely regulate the amount of cholesterol in the blood by manipulation of the diet alone. Most medical authorities agree, however, that some modification in eating habits can help to reduce high blood cholesterol levels and lower the incidence of cardiovascular disease. Though advising against becoming a "diet nut," the recommendations are to reduce high cholesterol food intake and keep the total diet low in calories by curtailing "high calorie foods." A low calorie diet is usually a low cholesterol diet. (Recommended dietary practices are presented in Chapter 7.) Other factors that tend to raise the level of cholesterol and other fatty acids in the blood are prolonged stress, smoking, and obesity. While it may cause an immediate rise because of demands for energy, exercise has been proven to reduce residual levels.

Overweight and diabetes. Statistics show that overweight people run 2 to 3 times the risk of heart attack that those of normal weight do—in addition to having a shorter life expectancy. This is because the over-weight person has a greater chance of having high blood pressure, high blood cholesterol, and diabetes—all direct risk factors in heart disease. The overweight person who develops or has diabetes is especially vulner-able because of the high levels of blood cholesterol and other fats that a diabetic condition builds up. *Diabetes* is a failure of the body to produce enough insulin to convert sugar and starches into energy. The result is a build up of excess sugar in the blood which in turn is associated with a rise in blood cholesterol level and other fats, thus leading to an atherosclerotic condition. The diabetic problem is shown in figures which report that over 3 million Americans have diabetes; only half of them are aware that they have it. For the control of diabetes, a variety of treatments including diet, weight control, exercise, insulin injections, and drugs have proven effective. Regular medical checkups are essential for early detection of the disease.

The magnitude of the overweight problem is reflected in statistics which report that over one-fourth of our population is significantly overweight. More shocking is that one-third to one-half of American children are reported to be obese—many extremely so. Since the amount of body fat, rather than actual weight, is considered a more significant factor in relation to blood cholesterol level, these figures may be either overstating or understating the problem. Body fat weighs less than muscle mass. Hence, a person may be within established norms for body weight, but still in excess of recommended levels of body fat. On the other hand, since most weight tables upon which these statistics are based do not reflect muscle-fat proportion, many persons may be "over-weight" but within recommended limits in terms of body fat. This will be explained further in the next chapter.

The causes of overweight and obesity are many and will be discussed in detail in Chapter 7. In only a small minority of cases, however, is it caused by an organic problem. In most instances, a properly supervised diet combined with exercise is sufficient to bring a person within his normal weight range or proper level of body fat. Unfortunately, only 50 to 60 percent of American households have diets which fully meet the standards of nutrition set by the National Research Council. Least likely to be eating the recommended kinds and allowances of food are the thirteen- to fifteen-year-old age group, with teenage girls the worst offenders. Overweight and obesity generally start early in life and are related to family patterns.

High blood pressure. High blood pressure or hypertension is an ex-cessively high and constant pressure placed on the walls of the arteries as the heart pumps blood through them. When prolonged, it can result in damage to the heart, kidneys, or other organs. It is a major risk factor in heart disease. Sometimes high blood pressure is caused by some other disease or emotional anxiety. When the cause is removed, the hypertension disappears. The exact cause of the most common kind of hypertension, however, is unknown. Far more is known about treating it. Methods of treatment which have been most successful include drugs, diet, avoiding of stress situations, and elimination of smoking.

Cigarette smoking. Much has been said about the relationship between cigarette smoking and lung cancer, but present research also shows a definite relationship between smoking and heart attacks and other circulatory diseases. The American Heart Association reports that the death rate from heart attacks among men smokers ranges from 50 to 100 percent higher than among nonsmokers, depending upon age and amount smoked. Though only half of all heart attack deaths are sudden (defined as occuring within fifteen minutes), among heavy smokers (over a pack a day), sudden death is two-thirds. On the other hand, smokers who have been able to quit the habit have a death rate nearly as low as nonsmokers.

The major effects of cigarettes on the cardiovascular system come from their nicotine content. Nicotine is a deadly poison. Injected into the blood stream, less than a drop is fatal. Taken in small amounts, as in cigarette smoke, both the heart rate and blood pressure are immediately increased by its effect on the nervous system and adrenal glands. With respect to sudden heart attack deaths, one of the common causes is a condition called *ventricular fibrillation* in which the heart rhythm goes wildly out of control, causing the heart to vibrate rather than contract. By its effect on certain nerve fibers in the heart, nicotine has been found to lower the threshold at which ventricular fibrillation occurs.

Nicotine stimulation of the nervous and glandular systems also tends to release hormones which, in turn, release fatty acids into the blood stream. Studies at the Philadelphia General Hospital in 1970 found this increase to occur within ten minutes after people smoked just two cigarettes. The free fatty acids then settle on the linings of the arteries and are believed to be a contributory cause of atherosclerosis.

Nicotine has a further effect on the circulatory system by constricting the capillaries and arteries. Blood flow is reduced in both the coronary (heart) and peripheral arteries (arms and legs). Skin temperature is lowered in both the fingers and toes. In persons with unusual sensitivity to nicotine, spasms may occur in the smaller arteries. Buerger's disease, a circulatory deficiency in the fingers and toes which often leads to the development of blood clots, gangrene, and the ultimate need for amputation, is a disease almost exclusively found in smokers. The condition almost always disappears when patients stop smoking and returns if they start again. For persons with atherosclerosis in the arms and legs, further narrowing of the arteries by smoking only increases the risk of gangrene, amputation, and death. Smoking studies have also shown that cigarette smoking speeds up the rate of internal blood clotting and makes clots tougher, further increasing the likelihood of coronary attacks, strokes, and pulmonary embolisms.

Physical inactivity. One of the major characteristics of our present-day society is the decreased amount of physical activity required to perform the tasks of normal living. Many studies have shown that men who lead sedentary or physically inactive lives suffer more heart attacks than those who are more physically active. In a study of over 31,000 London bus drivers and conductors, heart attacks were twice as common among drivers as among conductors who moved about and climbed the stairs of the double-decker buses. Similar results were found in studies comparing British postal clerks and mailmen, army officers and enlisted men in India (heart attacks 23 times more frequent among officers), South Dakota farmers and city dwellers, Swiss mountain villagers and townspeople, and a study of railway clerks, switchmen, and section hands. The railway employee study is significant in that the amount of physical activity in each job is fairly constant and, because of seniority, employees do little shifting from one type of job to another.

Follow-up studies of Cambridge graduates and of American college football players have also shown a relationship between the amount of physical activity and the incidence of heart attack. Cambridge graduates who had been active in sports had a lower mortality rate from cardiovascular disease than those who had not been active in sports while in college. The importance of maintaining a level of vigorous physical activity thoughout life was demonstrated in the study of college football players. Of the 335 players whose health records were checked, the cause of death was determined in 87. Twenty-five were found to be from coronary heart disease. Not one player who had maintained a vigorous exercise program was in the coronary group. The benefits of exercise to the cardiovascular system and specific programs for maintaining a vigorous level of physical activity throughout life are covered in Chapters 3 and 5.

other risk factors

Among the other risk factors associated with cardiovascular disease and heart attacks are low vital capacity, degree of alcohol consumption, a family history of heart attacks in middle-age, and having a spouse with coronary heart disease.

Low vital air capacity. Vital air capacity refers to the capacity of the lungs to take in and expel air. The latest data from the Framingham study shows that the higher a person's measured vital capacity (in liters), the lower his risk of coronary disease. Since the amount of oxygen the blood can deliver to meet the activity needs of the body and heart depends upon how much the lungs can supply, the vital capacity–heart attack relationship can easily be seen. While on the one hand vigorous endurance-type exercise can increase and preserve vital capacity, cigarette smoking has the most detrimental effect on it. A summary of the damage caused by smoking is presented in the section on lung diseases.

Degree of alcohol consumption. Basically, alcohol is a depressant—affecting the entire nervous system. The apparent stimulation of the first few drinks comes from the dulling of the higher brain centers which control thought, attention, memory, judgment, and inhibitions, as well as worry and anxiety. Similarly, the heart rate is initially increased by the depression of the vagus nerve which normally acts to inhibit or slow down the heart rate and keep it in check. The relationship between alcohol consumption and heart disease is now well established. Hospital studies have shown evidence of heart damage from alcohol. When alcohol has been removed from congestive heart patients, their condition improved. The condition returned when the patient returned to alcohol, often resulting in fatal congestive heart failure.

Recent experiments by Dr. Melvin H. Knisely at the Medical University of South Carolina have also shed new light on the destructive effect of alcohol on the tissues of the heart and other organs. Though the process is still not completely understood, alcohol seems to produce a substance which coats the red blood cells (oxygen-carrying cells), making them stick together and plug up many capillaries. The process has been called *blood sludging* and is created in varying degrees depending upon the level of alcohol concentration in the blood. The primary effect is to decrease or even cut off the oxygen supply to the tissues, depriving some of the cells of the oxygen they need to survive and causing them to die. The difference between the amount of tissue loss between the occasional, moderate, and heavy drinker appears to be only a matter of degree.

The cell destruction caused by blood sludging explains not only the damage to heart fibers found in heavy drinkers, but also the damage found in other organs. Autopsies on long-time alcoholics have revealed a number of small areas in the brain completely shrunken up—an apparent reason for the characteristic slow wittedness and poor judgment of these unfortunate persons. Blood sludging is also the reason that cirrhosis or hardening of the liver occurs 8 times more frequently in alcoholics than nonalcoholics. The effect on the brain, however, is most critical because, unlike the liver, it is unable to produce new functioning cells. At any rate, the sludging effect on the heart and other organs now helps to explain why life insurance statistics from age 25 show an additional life span of 39.5 years for heavy and moderate drinkers compared to 44 years for abstainers.

Adding to the effect of alcohol on the heart is both the number of calories it contains and the fact that people tend to eat more when drinking. Both factors can lead to overweight and the associated risks to the heart. As a food, of course, alcohol has no real value since it contains no fats, carbohydrates, and only minute amounts of protein and minerals. Indirectly, however, it has a food value that can be a major factor in weight gain. Its caloric value of seven calories per gram is exceeded only by fat. Through oxidation (combining with oxygen), the alcohol is converted into fats which are then stored with other fat deposits throughout the body. An added source of calories are the ingredients (sugar, flavorings, etc.) with which the alcohol is mixed for taste and coloring. Other factors are the stimulating effect on digestive juices and the appetite when taken in small and moderate amounts and the tendency to "nibble" and overeat while enjoying its relaxing effects. The low weight and malnutrition often common among alcoholics comes from their their loss of appetite for anything but alcohol. Lack of thiamine, a B vitamin, often causes beri beri—a disease affecting the nerves and circulatory system. Symptoms include muscle weakness and a swelling of the legs and heart caused by fluid retention. Congestive heart failure is the common result.

Family history and living habits. An hereditary tendency to heart disease is considered a definite risk factor. The chances are that if you had a close relative who died in middle age from heart disease, a tendency to the disease runs in your family. Though tendency does not make certainty, it should make a person more conscious of reducing other risk factors over which he has greater control.

That the tendency toward heart disease in a family may be more closely linked to similarity of living patterns than heredity was brought out in a study some years ago of adult males living in Ireland and their brothers who had migrated to America and were living in Boston. The Irish-Americans were found to have a higher incidence of coronary heart disease. Though their hereditary background was the same, the difference was in their living habits. The Irish were more physically active and weighed less, even though they consumed more calories than their brothers in Boston.

The greater risk of heart attack among those married to coronary heart disease patients again seems to indicate the importance of total living habits, since these are usually shared in a family.

other health and fitness problems

The emphasis so far has been on the major problem of cardiovascular disease and its relationship to the ignorance or neglect of the multiple facets of fitness. Cardiovascular disease is the chief reason that our life expectancy has not significantly increased during the last twenty years. Attempting to reduce this problem is reason enough for emphasizing the need for a broader approach to physical fitness and developing sound habits relating to it. Yet, there are other problems.

lung disease

COPD. The prevalence and death rate from *Chronic Obstructive Pulmonary Disease* (COPD) has increased spectacularly in recent years. COPD is a term used to describe conditions of chronic bronchitis, asthma, and emphysema which result in a persistent obstruction of bronchial airflow. The *bronchi* are the tubular air passageways which enter into each lung from the *trachea* (tubular passageway in the throat) and branch out many times until they are as small as capillaries. They form the network for delivering oxygen and removing carbon dioxide from the blood in every part of the lung. When any of its many branches becomes damaged, air flow is restricted. In emphysema, there are destructive changes in the smallest branches of the bronchi, tiny air sacs called *alveoli,* preventing oxygen and carbon dioxide exchange. The damage is irreversible. In asthma, the trachea and bronchi are over-responsive to various stimuli, causing a narrowing of the airways and difficulty in breathing. Excessive mucus secretion in the bronchi and a chronic sputum-producing cough are characteristic of chronic bronchitis. Because these conditions are sometimes so interrelated that they are difficult to distinquish precisely and because they share the common symptom of persistent airway obstruction, they are collectively called Chronic Obstructive Pulmonary Disease or COPD.

Despite the greater incidence of COPD in men than women, in urban than rural population, in black than white males, and in those exposed to occupational irritants, the relationship of smoking habits to COPD has been most impressive. Most COPD patients show a history of heavy smoking. According to the National Center for Health, the death rate from chronic bronchitis and emphysema (which studies in London and Chicago concluded were essentially the same disease) increased 300 percent between 1957 and 1968, with cigarette smoking playing a major role. Prechronic bronchitis symptoms of coughing, sputum accumulation, or a combination of the two are consistently found more frequently among cigarette smokers than nonsmokers. In addition, chronic bronchitis

—the usual forerunner to emphysema—is unusual in nonsmokers. Other early signs of approaching emphysema that are more frequently found among smokers than nonsmokers are shortness of breath upon exertion (*exertional dyspnea*) and reduced ventilary function (low vital capacity and forced expiration volume, caused by airway obstruction). The relationship between cigarette smoking and emphysema is so strong, in fact, that Dr. Charles Fletcher of Britain's Royal College of Physicians told a 1970 San Diego Smoking Clinic audience he considered emphysema "almost exclusively a smoker's disease." Though air pollution and occupational fumes play a part in aggravating COPD, they are not considered to play a major role.

The lung damage leading to COPD comes from the accumulation of cigarette tar on the linings of the trachea, bronchi, and tissues of the lungs. Tar refers to the concentration of smoke gases, liquids, and solids which forms a coating on the lining of these organs. Its primary effect is to slow down and, in time, even stop the self-cleansing action of the cilia in the respiratory tract. The *cilia* are tiny, hairlike tissue projections found in the lining of the air passageways. They constantly vibrate in a wave-like motion to remove mucus and other foreign particles from the respiratory tract. When these tiny tissue projections lose their "sweeping" effectiveness, smoke particles, dust, tars, smog chemicals, bacteria, and other harmful materials begin to accumulate, causing irritation and establishing a foothold for disease.

The vocal cords and membrane linings of the air passageway become thick and heavy. The bronchial mucus glands become hyperactive and an abnormal amount of mucus is produced. The mucus, or phlegm, can only be raised by coughing which further irritates the linings and causes more damage. "Smoker's cough" is usually a prelude to the development of chronic bronchitis. Infection and accumulation of harmful materials in the tiny bronchioles ("twigs" on the bronchial tree) and in the alveoli where the actual exchange of oxygen and air takes place, may ultimately lead to emphysema and a permanent loss of breathing capacity. Its most common manifestation are ruptured air sacs and inelastic bronchioles.

Normally, there are some 300 million tiny air sacs in the average pair of lungs. Their walls are so thin that the oxygen–carbon dioxide exchange takes place through capillaries which form a mesh so thin that the alveoli are virtually covered with blood. When the alveoli walls break down, larger air pockets are formed (in advanced cases, the lungs look like an old-fashioned bath sponge). Thus, there is less surface area for the oxygen–carbon dioxide exchange; the body can no longer take in all the oxygen it needs or get rid of all the waste gases effectively. Loss of elasticity in the bronchioles leading to the air sacs causes them to collapse. They can then be forced open only with considerable effort.

With the loss of lung capacity comes difficult and labored breathing. In addition to its effect on breathing, emphysema is a contributory factor in heart disease, since the heart must work harder in order to provide sufficient oxygen to the body.

Emphysema is irreversible. Proceding from a chronic cough to the worsening of dyspnea, the advanced COPD patient begins to show a loss of weight and appetite from the effort required for simply eating. With too little oxygen and too much carbon dioxide reaching the brain, loss of mental function, memory, and judgment occur. Headaches, drowsiness, apathy, confusion, tremors and twitching of the extremities, and— ultimately—coma and death follow.

Treatment of COPD patients is aimed primarily at preserving whatever lung function there is and helping the person adapt to the limitations imposed by the damage. Exercise cannot prevent or correct lung damage, though it may delay it. For some patients, exercise can even be dangerous. Breathing exercises and a gradual exercise program to improve general muscle tone and increase work tolerance, however, may be prescribed. For the majority of patients with mild or early COPD, the avoidance of cigarette smoking may be the only treatment needed, according to a COPD manual for physicians published by the National Tuberculosis and Respiratory Association. Avoiding inhalation of occupational dusts and fumes, along with lessening exposure to cold air, are other nonmedical recommendations.

Lung cancer. Besides the ultimate effect on respiration and the heart, there is growing laboratory evidence relating tar accumulation and certain of its substances to the development of lung cancer. Early in 1970, the American Cancer Society reported that in experiments at the Veterans Administration Hospital in East Orange, N. J., researchers were for the first time able to produce lung cancer in experimental animals trained to smoke as humans do. Prior cancer producing experiments involved only the painting of tars on animal tissues, a method often criticized by independent researchers skeptical about the dangers of smoking.

Other studies at the East Orange hospital found that tar accumulation was able to produce precancerous changes in the cells of the trachea and bronchi. Such changes are termed *noninvasive* cancer—changes that are similar to cancer cells but which have not as yet penetrated the surface membrane, yet which are an almost certain preliminary to cancer. Significantly, studies of ex-smokers showed that in those who had precancerous lung cells at the time they stopped and who didn't develop cancer, the abnormal cells disappeared over a period of time and were replaced by healthy cells. Further studies showed that such noninvasive cancer could be detected by examining sputum coughed up. The discovery offers the possibility of an excellent screening device for a disease which even with surgery offers less than a 30 percent chance of recovery, according to some experts, and only 5 percent according to the American Cancer Society. The fact that the precancerous condition can be detected so easily and reversed merely by quitting cigarettes should encourage smokers who hesitate to stop out of the false notion that "the damage has already been done."

psychosomatic and nervous ailments

Psychosomatic ailments are physical disorders which result from psychological or emotional problems rather than from specific body diseases. They are the result of body chemistry changes brought about by prolonged emotional stress. Physical and emotional fatigue, headaches, backaches, colitis, ulcers, hypertension, and many other physical ailments may be the result of excessive nervous tension. In a study of outpatients at the Yale University Clinic, it was found that 76 percent of the patients had illnesses caused by emotional stress. There is some evidence to indicate that lack of exercise makes a person a more likely candidate for nervous tension and the psychosomatic disorders which often accompany it.

Recent years have also seen a tremendous rise in the number of persons suffering from nervous breakdowns and mental illness. Much of this rise can be attributed to the increased pace of living; the increased tensions of local, national, and international affairs; the inability to cope with rapid social changes; and the inability to handle private emotional problems. Treatment necessarily involves competent psychological and psychiatric counseling and medical care. Drugs and electroshock therapy are often used to reduce tension and prepare the patient for guided insight into the underlying causes of his problem. But proper diet, exercise, rest, and relaxation therapy are also used to prevent and reduce the build up of tensions in mental hospital patients. We can imply from this the value of these practices for those whose daily tensions fortunately have not reached the point of requiring hospital care. A further discussion and some specific methods of dealing with stress and tension are covered in Chapter 8.

drug abuse

Experimentation and dependence on drugs has increased at what has been called an epidemic rate since 1965. Many of those in mental hospitals are there because of drug abuse. While once it was considered a problem only in the ghetto districts, today it is present in all classes of our society and among the young and old alike. To some hopeful observers, the epidemic has peaked and is beginning to decline. If so, much of the credit must go to the widespread publicity given to the dangers of drug abuse, to governmental efforts in controlling drug traffic, and to the many outstanding drug education and rehabilitation programs established nationwide in which the testimony of former drug addicts has played a major role. Because of the impossibility of completely cutting off the flow of drugs, however, and because of the socio-psychological factors involved in drug use, the problem seems destined to remain for some years.

It is beyond the scope of this book to go into a detailed treatment of drugs—their nature and effects and the socio-psychological factors involved in their use. As with smoking and drinking, however, drug abuse is recognized as one of the negative factors in the development of optimum health and fitness. In addition, because many of the reasons given for the experimental and social use of drugs reflect emotional and social voids which others have filled through participation in a wide variety of physical, recreational, and athletic activities, some further comments seem appropriate and consistent with the general theme of creating a broader concept of fitness.

First, it should be recognized that as with alcohol and its use, not all drugs or drug users can be lumped together into one category. There are enormous differences between the various types of drugs, the reasons for their use, and those who use them once or twice and those who make them a way of life. Not all users can be considered "abusers." Secondly, though discussions of drug abuse usually center primarily on the young, it should be recognized that the problem does not involve all young people, nor is it restricted to youth alone.

Motivations for drug use. Though changing cultural values and patterns of living have had an influence, many of the reasons for the experimental and occasional use of drugs are really not new, complex, or abnormal. They often reflect normal desires, are typical of youth in any generation, and are similar to reasons for experimenting with alcohol, tobacco, or sex. According to users themselves, most are introduced to drugs by their friends, or peer group, and begin out of simple curiosity. To others, it's the fashionable thing to do—they want to be part of the "in" group. Behind this motivation is the natural desire for group acceptance and a sense of belonging. To many, the appeal is a natural sense of adventure, daring, and excitement—similar to that which motivates high speed driving and certain forms of athletic competition. Basically, to these persons, drugs are simply a source of "fun" or "kicks." Some may have normal difficulties in communication, shyness, establishing close relationships, or be troubled with frustrated sex drives and see joining the drug crowd as a solution. To the more intellectually oriented, marijuana and the so-called mind-expanding drugs offer an opportunity for deeper personal, religious, or philosophical insights. A few turn to drugs out of open dissatisfaction with their total life pattern, as a crutch to cope with their personal problems, or out of rebellion against parents or a society they see as materialistic, hypocritical, or repressive of personal freedom and expression.

Development. For many, drug experimentation is a one time experience only. Their curiosity satisfied, the effects really not that pleasant—perhaps, even unpleasant—they have no further motivation to continue. Others, suffering no apparent ill effects, continue much in the fashion of the occasional drinker. They may feel the warnings against drugs have been exaggerated. Many honestly consider controlled drug use no different from the socially accepted use of alcohol and add an antilegalistic motive for continuing—to combat what they consider stupid, ungrounded laws against drug use. Important considerations and differences are discounted (equal potential danger of alcohol addiction, relative ease of equivalent intoxicating effects from drugs, easier immediate availability and tendency towards frequent use, exposure to a greater variety of drugs). Some, feeling that they have been lied to, come to doubt all "authoritive" statements about drugs. Falsely confident, they may go on to become chronic abusers.

The chronic abuser. It is among the chronic abusers that the greatest danger lies and where the motivations become more complex. Even here, however, the underlying psychopathological problems are not unique and are in many ways similar to those leading to chronic alcoholism. Though some young people who have become "hooked" on drugs have been described as "perfectly normal, well-adjusted kids," the majority of regular users seem to be those whose original motivations came from an abnormal desire for "kicks" or from deep feelings of insecurity, dissatisfaction, or rebellion. The progress from experimental to occasional and, finally, to regular use is usually quite rapid in these persons. Characteristic of many users with deep-rooted psychopathological problems, according to former drug addicts themselves, is an inability or unwillingness to tolerate stress or frustration and deep feelings of depression and inferiority. Drugs offer an escape from the need to plan, struggle, endure, and cope with problems. Some are bored and unable to find enjoyment from ordinary existence, either from within themselves or from without—drugs provide the missing spark. Others find communication and deep interpersonal relationships so difficult that drugs become the only way to make them easy and gain a semblance of deep intimacy. In actuality, the result is only a deeper turning inward. The drug-induced silent communication of thought and intimacy, in reality, turns out to be only the creation of their own minds and projection of their own feelings. The hurt, dissatisfaction, anger, and rebellion of others is so intense that drugs become for them their major weapon for striking back at parents, the establishment, or society.

Former drug addicts have been almost unanimous in citing two common factors present in almost all of them. One was that in all something spiritual or emotional was missing in their lives that led them to seek fulfillment through drugs. The second was that, even though they were vaguely aware of the potential danger from drug use, none believed they would be harmed or, certainly, that they would ever become "hooked." Equally unanimous have been their statements of gaining only temporary pleasure and of the ultimate realization that the benefits gained were unreal—only chemically induced.

Drugs and you. Though society may be partly responsible for creating a climate for drug abuse, in the final analysis the decision to use or not to use drugs remains an individual one. It is a free choice, based on your own individual values of what you really want out of life and what you consider the best way of attaining it. Certainly there are many who without the aid of drugs have satisfied curiosity, found group acceptance, excitement, and deeper insights into themselves and others. Also, since no one is free from personal and social difficulties, feelings of dissatisfaction, and tendencies toward rebellion, there are obviously other avenues for the solution of these problems.

Being sufficiently informed about drugs and their risks is the first requisite for making a decision about them. Almost every school and community today has some form of drug education program with materials that give the basic facts about drugs and discuss some of the factors leading to their use, as presently known. Several excellent sources are listed in the bibliography at the end of this chapter. In addition, try to attend a panel discussion, lecture, or audio-visual presentation involving former drug addicts. Listen to what they say.

The second requisite for making a decision about drug use is a careful examination of the reasons you might have for wanting to use them, and a consideration of alternate ways in which the same benefits might be gained. Former drug users have stressed that the use of drugs was a reflection of something emotionally or spiritually missing in their lives. Not many persons will deny the validity of such reasons for turning to drugs. Most question only the necessity of drugs to achieve what is being sought and the ability of drugs to provide any lasting benefits. Being part of a group, having a sense of belonging and recognition are natural desires. So too is the desire for adventure, excitement, pleasure, self-awareness, and even the desire to avoid or overcome difficulties. Dissatisfaction and concern with the defects of our society and the emptiness of materialism, along with the quest for personal fulfillment and deep interpersonal relationships are also perfectly valid feelings shared by many. The difference is that the nonuser has found alternate routes for achieving these same goals. It is in finding these alternate routes that the ultimate solution to the problem of drug abuse appears to lie.

For many, one alternate route is found in the challenge, sense of accomplishment, and warm interpersonal relationships that come from skill development and whole-hearted participation in the wide variety of physical, recreational, and friendly competitive athletic activities discussed in the last chapter of this book. The increased physical vigor and sense of total well-being that can be gained from these and the other activities, principles, and techniques that will be presented are additional routes well worth exploring.

low back problems

Problems involving the lower back have become serious for many middle-aged Americans. It has been estimated that the number of people suffering from low back pain is second only to those suffering from heart disease. In a study of Los Angeles County firemen retired on disability, heart and back injuries accounted for approximately 75 percent of the total pension cost. There are several causes of low back pain, but many can be attributed to lack of physical activity and proper exercise. Weaknesses of the ligaments attached to the vertebrae may be a cause of instability of the spine. Weakness of muscles in the abdominal and lower back areas, loss of flexibility in the low back and hamstring muscles, poor posture, poor body mechanics in performing ordinary tasks, nervous tension, and general fatigue are other causes. In some cases, back pain may be caused by disease, pathological conditions, or hereditary defects. Exercise, good habits of posture and body mechanics, and techniques of relaxation can do much to eliminate or relieve this nagging disability.

physical deterioration

In addition to the problems already discussed, there is evidence of a general physical deterioration which is associated with our present-day patterns of living and neglect of basic fitness habits. The age in which we live seems to have bred underexercised, overfed individuals—young and old—who are fat, flabby, and weak in the upper trunk, back, abdominal areas, and feet. Lack of physical activity has been shown also to be related to the prevalence of varicose veins, hemorrhoids, and muscle atrophy. Some years ago, Thomas K. Cureton, Ph.D., a physical fitness specialist at the University of Illinois, said that most Americans begin to display middle-age characteristics at age twenty-six! Some wonder today if this was not an understatement. Many of our young men continue to be judged unfit for military service for a variety of medical reasons. Of those accepted, many can't swim, never have engaged in vigorous sports or other physical activity, and have to be placed in special conditioning battalions and on special diets.

changing living patterns

It should be clear that despite advances in medicine and technology which have made life easier for us, many serious health and fitness problems exist that are a threat to us as individuals and as a nation. Laborsaving devices in the home and at work, together with the automobile, have taken much of the previously necessary physical activity out of our normal living activities. The increased time spent sitting both on the job and in the home describes what is called the *sedentary life.* A general high standard of living and affluence have drastically changed our eating and other living habits. Increased leisure time, rather than becoming a blessing, has for many resulted only in a continuation of sedentary living, overeating, increased tension, and unsound health practices. There is no question that the major health problems today have to do with our patterns of living, and this is evident not only in the United States, but in the leading industrial nations of the world.*

A 1970 United Nations report states clearly that man has lengthened his life-span as much as he can by fighting contagious disease. Heart and lung ailments, cancers, strokes, auto accidents, and suicides are listed among the major causes of death today. The report states that to increase the life-span any further, a change in our way of living must occur. We must introduce new factors capable of preventing death from a different range of diseases.

Certainly no one wants to return to the hard life of the pretechnological age when work on the farm and in the factory amounted to from twelve to sixteen hours a day. Neither do we want to return to a time when the automobile was a luxury and recreation was at a premium. The age of automation is here and will continue to make life easier and abundant for all. Consistent with today's greater emphasis on personal freedom and responsibility, as opposed to the authoritarianism and harsh discipline of the past, what is needed is a greater understanding and a new adjustment of our total living habits to the requirements of this new age. Though many other factors are involved, we do know enough now to say that developing sound knowledge and habits of fitness can be a beneficial start. Knowledge and habits relating to exercise, diet, rest, relaxation, smoking, drinking, drugs, and sport and recreational activities can make a difference. We can't wait for a pill to solve our health and fitness problems.

Young and old alike must learn about how their body functions and about their health, physical activity, and recreational needs. They must learn how and how much to exercise and what kinds of exercise are best for various purposes. There must be a sufficient emphasis on endurance-type activities and on sport and recreational activities that will meet lifetime physical, emotional, and social needs. They must know how to eat, what foods to avoid, and about body weight and body fat. They have to be informed about smoking, drinking, and drug use—know the effects and have some insight into the reasons for their use. In short, they must have a broad concept of fitness, know the how and why, and develop lifelong patterns of fitness.

In this book, we have and will be focusing on each of these multiple facets of fitness to give the basic understanding, principles, and skills needed to reach and maintain a minimal level of fitness throughout life. The emphasis will be on learning specific principles and techniques, based on sound knowledge, that will:

1. Improve your health and well-being
2. Provide you with a defense against diseases and disabilities resulting from improper care of the body
3. Give you the ability to perform your daily tasks without undue fatigue, and with enough energy left over to meet extra demands and emergencies and to enjoy pleasurable recreational activities
4. Help you to combat the mental and emotional stresses of present-day living and their tendency to encourage unsound health practices
5. Give you a new self-image and an attractive appearance, within the limits of inherited body characteristics and permanent disabilities
6. Provide you with a foundation for engaging in pleasurable sport and recreational activities
7. Help you to add zest to your living and enjoy a full, fruitful life

components of fitness to be developed

The components of fitness are the qualities or abilities of which it is made. Physical fitness obviously implies good health or freedom from disease. One of the most important performance components of physical fitness is general endurance or working capacity—the ability to work for a long time without undue fatigue. This is also often referred to as *circulorespiratory endurance, cardio-respiratory endurance,* or "wind", and has to do primarily with the condition of the heart, lungs, and entire cardiovascular system. It is evident that weakness in this area is one of our most serious problems today. Included as a part of physical fitness is enough muscular strength, muscular endurance, flexibility, agility, balance, and coordination for safe and efficient movement in work or at play. These elements are collectively referred to as *motor fitness*. Proper body proportions (weight, girth, body fat, and lean muscle mass), in accordance with inherited body characteristics, is another component important not only for good health, but for an attractive appearance and a good self-image. Good posture and the use of proper body mechanics in performing ordinary tasks such as lifting, pulling, pushing, and carrying, are components which, if ignored, can damage appearance and lead to functional disability. Finally, the ability to relax, including adequate rest and recreation, and ability in a wide range of sport and recreational activities are other components of a broad concept of fitness.

In addition to these positive components of fitness, consideration must also be given to negative components, such as smoking, drinking, and the use of drugs. Certainly, attitudes toward these practices vary. They are determined by those of family and friends, by personal experiences, and by knowledge about their nature and effects. In addition, the decision to smoke, drink, or use drugs is recognized as an individual choice based on a person's own value system. Whatever the decision, however, it should be based on a careful study of the facts, some insight into the reasons for such practice, and a consideration of whether the risks involved are worth the benefits gained. With respect to drug use, unless present laws are changed, such consideration would also include the legal consequences involved. An examination of the reasons why others have either quit or abstained from these practices, and the ways in which they have substituted for their supposed benefits, can also be helpful in making decisions. At any rate, it should be understood that those who do not smoke, drink, or use drugs will achieve a higher level of overall fitness than those who do. But even those who are unwilling or unable to abstain or quit can be helped by developing the positive components to the highest degree possible.

2

evaluating
your
fitness level

Before attempting to make any changes in your present level of fitness, we should first have some idea of where you stand in relation to the components of fitness presented. The information is essential to give you an accurate insight into your physical condition and potential. Only in this way can you accurately determine your needs and devise a sound and safe plan for meeting them. The evaluation will also provide a basis for measuring improvements following whatever action you take to raise your level of fitness (exercise program, dieting, giving up smoking, etc.). Unless you have an initial basis for later comparison, there is no way to measure objectively the success or failure of your efforts.

Evaluation items to be discussed and presented include some which will require the services of professional personnel and some that can be administered individually or with the assistance of another person. A Fitness Evaluation Summary Form in Appendix A can be used to record the results of your initial and later evaluations.

the medical examination

A thorough medical examination, including a chest X-ray and other laboratory tests, is an essential first step for a complete evaluation of your physical condition[1] Both the medical examination and tests of physical performance are measures of fitness along a wide scale, ranging from the absence of health at one extreme end to the vigor of a trained athlete at the other end.

Along with a review of your health history record, the medical examination is necessary also for safe guidance in any fitness program and for identifying areas that may need special attention. The nature and intensity of exercise programs must be adapted according to the findings of the physician. Limitations, if any, must be pointed out to avoid possible aggravation of conditions and even serious harm. Specific exercises may be prescribed for the improvement of certain conditions. Certain special precautions may be necessary in a weight reduction program. On the other hand, if no past or present limitations are found, you can proceed with no worry of serious ill effects.

importance of the annual medical examination

In terms of its overall importance in the broad area of fitness, the importance of having an annual medical examination—even when you "feel good"—should be stressed. Though faithful attention to the principles and programs discussed in this book will greatly reduce your chances of developing many premature serious illnesses or disabilities, there can be no absolute guarantee. Too many unknown factors are still present that can lead to serious disease. Most of these diseases are present for a long time and to a mild degree before outward symptoms appear to indicate that something is wrong. Only a thorough medical examination with associated laboratory tests can detect these diseases in their early stages. Such early detection means a greater chance for correction before the condition develops into a serious problem that can become costly, disabling, or even fatal.

components of the medical examination

The actual components of the examination vary according to the judgment of the physician and the person's age, sex, and health history. For checking the lungs and heart, a chest X-ray is considered mandatory, along with a check with a stethoscope and a blood pressure reading. For those over thirty, an electrocardiogram (heart tracing) is usually recommended—preferably under moderate stress rather than in the resting

---

[1] Condensed and adapted with permission of the author from *The Annual Medical Examination—What Should It Consist Of?* by Dan H. Eames, M. D., La Marque, Texas, 1967.

state, since these readings have proven more accurate in detecting abnormalities. Noted heart specialist, Dr. Paul Dudley White, recommends an EKG even at age twenty, so that it can be filed and used for later comparisons. Another recent recommendation by many physicians is a blood serum cholesterol count for all age groups, including young children.

Other basic laboratory tests considered essential are a urine analysis for screening a possible diabetic condition and a blood count to detect possible anemia, leukemia, or tip off an unsuspected infection in other parts of the body. *Anemia* is a condition in which the red corpusules in the blood are reduced or deficient in hemoglobin (oxygen-carrying cells). It can cause pallor, lack of energy, shortness of breath, and palpitation of the heart. A routine part of the examination is also an internal examination to detect possible mass growths. For women, an annual Pap smear (cancer smear) taken at the mouth of the womb (cervix) is considered mandatory, since the cervix is one of the most frequent sites of cancer development in the female. Because the test can detect changes one to two years before a cancerous stage is reached, cancer of the cervix could virtually be eliminated if every woman had an annual Pap smear.

Other specialized tests may be given at the discretion of the physician and on the basis of the health history and preliminary findings. Because venereal disease has become a major health problem among young people today, mention should be made of the VDRL blood test. Many outward symptoms of venereal disease may go unrecognized and disappear, but the infectious disease remains within the body and continues to work its harmful effects. Years later, when outward symptoms again appear, the damage may be irreversible. Only a blood test can detect the presence of venereal disease in the absence of outward symptoms. A VDRL blood test is advisable, therefore, whenever there is any reason to suspect possible VD infection from any cause.

body build

Having some idea of your basic body build is helpful because definite relationships have been shown between a person's body type and his health, susceptibility to various illnesses, and physical capacities. In addition, the extent to which you can make changes in your body type is limited. It is therefore important both as a guide and for an accurate self-image. The procedure for determining a person's basic body type in terms of inherited characteristics of bone, muscle, and fat distribution is

Intro para

called *somatotyping*. It was developed by Dr. William B. Sheldon and his associates at Columbia University many years ago.

major body types

Sheldon, who has made a lifelong study of the human physique, has classified the human body into three major types: *endomorph, mesomorph* and *ectomorph.*

Intro. para

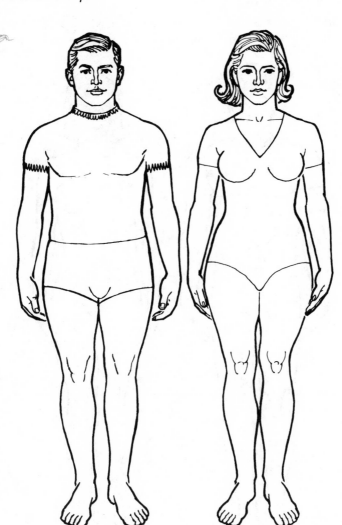

FIG. 1.   Endomorphic Body Build

Endomorph. This body type is characterized by a predominance of fat tissue (not necessarily obese), with a large body, wide abdominal area and hips, and relatively short arms and legs. Weight is concentrated in the center of the body, giving a "blocky" or in the extreme a "pear-shaped" appearance in relation to the upper trunk. Little bone and muscle definition creates a general appearance of smoothness and roundness. Upper arms and legs are "hammy." Bones and joints are relatively small compared to body size.

Mesomorph. In men this build is characterized by a solid muscular build; muscles, bones and joints are thick and heavy. This build has

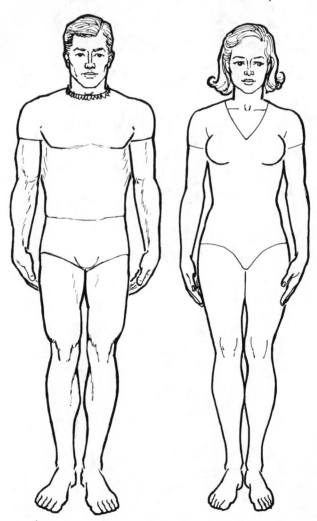

Fig. 2.   Mesomorphic Body Build

broad shoulders, a massive chest, slender and low waist, and broad hips. The upper trunk has a "V" shape, the buttocks are heavy, and the thighs large. In women, this is a solid build with muscles large and firm, though not as pronounced as in men because of the adipose or fat tissue covering which is normally greater in women than in men in all body types. There is good muscle in the arms, abdomen, buttocks, and legs with relatively thick and heavy bones and joints.

Ectomorph. This build gives a linear appearance, with a frail and delicate bone structure, thin and stringy muscles and usually very little

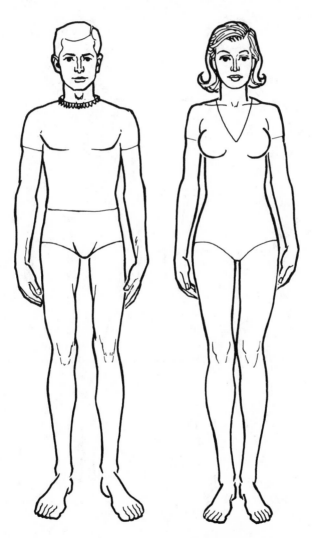

FIG. 3. Ectomorphic Body Build

fat. A relatively small and narrow chest and a shorter trunk in relation to the arms and legs is also characteristic.

other general classifications

Two other general classifications of body types are medial and ecto-medial.

Medial. This is characterized by medium build and muscle structure with no predominant physical features.

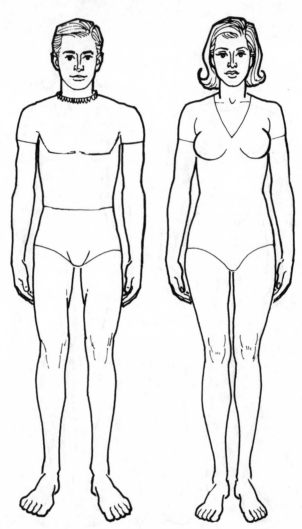

FIG. 4.   Medial Body Build

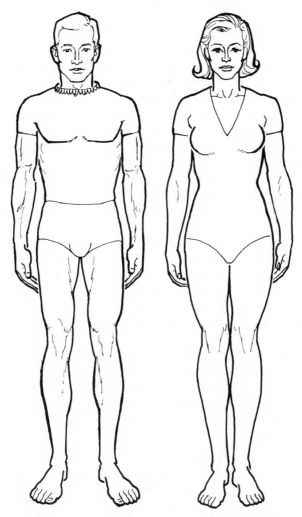

FIG. 5.   Ecto-Medial Body Build

Ecto-medial. This is the thin, wiry type with a very sharp muscular outline, though not of sufficient size and thickness to be considered mesomorphic.

significance of body build

Limits changes that can be made. The claims made by certain magazine and newspaper advertisements that certain diet and exercise techniques can change anyone into a supposedly "ideal" Mr. or Mrs. America are simply not possible. Neither is there any such thing as an "ideal" physique which all must strive for.

Basic body build is inherited. The length and thickness of muscle fibers are inherited characteristics that remain relatively fixed throughout life. Stocky types (endomorphs and mesomorphs) have a larger digestive tract per unit of body volume and larger internal organs than slimmer types (ectomorphs, ecto-medials). Because of this greater capacity to take in food and absorb nutrients, it is logical that endomorphs and mesomorphs tend to be heavier and have more body fat than ectomorphs. Exercise and diet can certainly bring about significant changes in muscle size and the amount of fatty tissue, but it is important to recognize that these changes will be limited by your basic inherited body build.

Everyone has to work with what he has, but through proper diet, exercise, and attention to other areas of fitness anyone can develop a reasonable level of fitness and bring out the "unique attractiveness" of his or her own particular body type. The goal, therefore, should be to develop your own particular body type to its best proportions. Physical attractiveness, after all, does not lie in having one body type rather than another. It is based on muscle development and fat distribution in proportion to the body frame and without excesses caused by unsound practices. Under- or overdevelopment of muscles, for example, distorts body proportions regardless of body type. This is why, to many persons, the overdevelopment of muscles by "body builders," even if "meso-morphic," appears as grotesque as do the muscles of an extreme ectomorph who has never bothered to develop them to their full potential. The same can be said about excesses in body fat caused by poor eating habits rather than constitutional factors.

Body type and health problems. Certain health problems and greater susceptibility to particular diseases are associated with body type. Predominant endomorphs have a continuing problem of weight control and must constantly be on guard to keep their weight within reasonable limits. According to Dr. Jean Mayer, who has made a life study of diet, nutrition, and problems of overweight at Harvard University, the prime requisite for obesity is at least "a moderate amount of endomorphy under normal nutritional conditions."[2] He points out that, although not all endomorphs become obese, the tendency is there unless "insufficient diet, great activity, disease, or voluntary weight control supervenes."

2 Jean Mayer, *Overweight* (Englewood Cliffs, N.J.: Prentice-Hall, Inc., 1968), p. 42. Dr. Mayer is considered one of the world's foremost authorities on nutrition. He has served on a number of United Nations Committees on Nutritional Requirements, and in 1970 was appointed by President Nixon to head a national effort aimed at improving the nation's nutritional habits.

Because of their greater tendency towards obesity, endomorphs run the risk of becoming more susceptible to circulatory disorders, kidney problems, liver ailments, and diabetes. For these reasons, the endomorph must be conscious of keeping his activity level high and his calorie intake low. Proper diet and exercise can help him to become a reasonably fit and lean person.

The mesomorph is usually very active, robust, and healthy in his younger years, but has a tendency toward excesses in eating, drinking, and smoking which must be curbed. Unless he remains active, he has a tendency to become overweight and is prone to develop obesity, especially in later years. Combined with his excesses, he can become susceptible to most of the same circulatory, kidney, and liver disorders that the endomorph faces. To feed his large body build, he needs to sustain good circulation by lots of heart-respiratory exercise.

Health problems most often associated with ectomorphic types include respiratory diseases such as tuberculosis, pneumonia, and common colds, and nervous ailments. Weak muscles, tendons, and ligaments make them prone to postural defects and injuries as well. Ectomorphs are also inclined to lack endurance and to be physically lethargic. They need to concentrate on building strength and endurance.

The ecto-medial is full of nervous energy. He is a fast mover who hates to sit down and is noted for fussing, worrying, and talking fast. Though he rarely gets fat, he is usually bothered by nerves, digestive ailments, and tension diseases. He is the classic "ulcer type." His greatest need is to get lots of rest, learn how to pace himself, and to relax.

The medial is usually very active, ambitious, and full of energy. He hates to be held back and usually presses very hard. In later years, he is inclined to develop a paunch, but usually stays lean in other areas. He is also prone to suffer from heart, stomach, and blood pressure disorders. Like the ecto-medial, he needs to get lots of rest, control his energy output, and learn how to relax.

Body type and physical capacity. Though individual differences in nutrition, training, and specific physical skills play an important part, your basic body build will determine to a large extent your chances of success in various physical activities. Mesomorphs are usually stronger and have more speed, agility, power, and endurance than either endomorphs or ectomorphs. As a group, they have almost unlimited athletic potential and usually enjoy vigorous, exciting activities. They have a particular advantage in body contact sports such as football and wrestling.

Endomorphs are stronger than ectomorphs (though not necessarily in strength per pound of body weight), but have less speed, agility, and endurance. Usually endomorphs are unsuccessful in activities that require support of the body weight such as gymnastics, and in activities that require speed, agility, and endurance. Their body weight is a handicap in performing most tests of physical performance and they can be expected to score less than mesomorphs and even ectomorphs in most areas. Their total strength and weight, however, can be an advantage in football (as a lineman), wrestling, and other contact sports, provided they have adequate movement. They can usually be successful in golf, archery, and other activities where agility and speed are not primary requirements. Because of his buoyancy, the endomorph can also be a very graceful swimmer.

Provided they have an adequate degree of muscular development, ectomorphs can be successful in activities requiring support of body weight (gymnastics) and in racket sports, track, basketball, and other nonbody contact sports. Since injuries are frequent in contact sports, these activities should generally be avoided. However, many predominant ectomorphs can be successful in football as ends or running backs if they have sufficient muscle development.

Medials and ecto-medials are most versatile and can generally be successful in most physical activities. If lacking sufficient height and weight, their primary limitations might be in contact sports because of the possibility of injury. In sports such as wrestling, boxing and judo, however, where competition is based on weight class, this danger is minimized. Selective placement in certain football positions (quarterback, running back, or defensive back) would minimize the limitations in this sport also.

determining your body type

Though certain objective criteria are used as a guide, all body type ratings are usually subjective and result in only a rough classification. Detailed procedures for rating body types have been developed by W. H. Sheldon, Dr. R. W. Parnell, and others. Though more accurate, they are too time consuming for our purposes. We want only to make you familiar with the major body types, and enable you to roughly determine your predominant characteristics, so that you can gain greater insight into physical potential and possible health problems.

By comparing your body lines with the pictures and characteristics given in Figs. 1 to 5, you should be able to identify your predominant body type or predominant combination. Endomorph, mesomorph, ectomorph, and ecto-medial classifications should be used if you show a very definite predominance of characteristics associated with any of these body types—regardless of height. If you have no predominant features, classify yourself as a medial. Combination body type classifications can be made if you show characteristics of two major types. A meso-ectomorph, for example, would be a person who is primarily an ecto-morph (long, thin bones, and small joints), but also well-muscled, as are many basketball players. Similarly, a meso-medial would be an average proportioned individual with a little above average muscular development.

posture analysis

Posture is a position assumed by the body. The position may be either a static or a dynamic one. *Static* positions are those in which the body is relatively inactive or immobile, such as when standing, sitting, or lying. *Dynamic* positions are those which involve movement, such as in walking, lifting, pushing, or pulling.

Good static and dynamic posture is important in order to have an attractive appearance, to avoid interference with internal functions, to prevent aches and strains, and to maximize mechanical efficiency. Later we will discuss further the basic principles and techniques involved in assuming various postures. At this point our concern is with analyzing normal standing posture, since it is basic to all other postures and usually reflects the way an individual habitually carries himself—not only when standing, but when moving about and performing other tasks. Our purpose is to see if any deviations from proper basic alignment, or the beginnings of serious defects, that can be detrimental to your appearance or mechanical efficiency can be detected. In cases where correction is possible, specific exercises can then be prescribed or referral made to a physical therapist, chiropractor, or qualified medical doctor for opinion and guidance.

causes of postural deviations

In some cases, poor body alignment may be caused by structural or skeletal damage stemming from birth, accident, or disease. Aside from surgery, there is usually little that can be done for these conditions. Many postural deviations, however, are simply the result of weak or unbalanced muscle development. In these cases, the chances of correction are greater unless, of course, the muscle weakness has been caused by a disease such as polio.

muscular development and posture

In the absence of structural defects, the skeletal frame is held in a balanced, upright position by the equal contracting force of opposing muscle groups. For example, contraction (shortening) of the muscles in the front of the neck pulls the head forward, while contraction of the muscles in the back of the neck pulls the head backward. In the normal resting state, each of these muscle groups is in a slight state of contraction known as *tonic contraction*. When each set of opposing muscles is equal in development and degree of tonic contraction, the head is held erect. Similarly, the shoulders are kept in alignment with the head by equal levels of tonic contraction in the chest and upper back muscles. The chest muscles exert force forward and downward, while the upper back muscles pull the shoulders upward and backward. Balanced tonic contraction of muscles which move the skeletal frame forward, backward, or laterally at the various joints is therefore essential for proper body alignment.

Over a period of time, habitually poor body carriage or certain repeated movements in work or play activities that overwork one set of opposing muscles at the expense of the other—combined with a lack of properly balanced exercises—can lead to unbalanced muscle development and postural deviations. An example of adaptation to a poor work posture over a long period of time is the round shoulderedness seen in many typists. Basketball players often develop the same problem from the repeated contraction of the chest muscles in dribbling and passing and the corresponding stretching of the upper back muscles. Deviations caused by unbalanced muscle development may ultimately result in permanent structural deviations. Unfortunately, the gradual development of these postural deviations often goes unnoticed until they become serious enough to be obvious or cause functional disability.

Basically, correcting deviations stemming from unbalanced muscular development involves exercises based on the simple principle of "Stretch the Short Side and Strengthen the Long Side." Chapter 6 goes into more detail concerning the nature, causes, and possibilities for correction of various defects and about preventive measures for avoiding them. Specific exercises for their correction are also included. Your knowledge of this information, along with the ability to recognize deviations, can be of value to you not only personally, but as a parent. Early detection and appropriate remedial action can save your children from what might be irreparable physiological and psychological harm.

making the analysis

The posture analysis can be made by either comparing a series of self-photographs against the Posture Analysis Form shown in Appendix A or by having another person check you against the form. In either case, clothing worn should be minimal or close fitting, such as a bathing suit or leotard, and the background solid and contrasting, with vertical and horizontal guidelines if possible. Care should be taken to stand naturally, with the heels together, and to avoid a posed or false posture. If pictures are taken, they can be used to make a more studied determination of your body type by adding a front view to the side and back views shown on the analysis form.

body proportions and body fat analysis

Various body measurements, including height, weight, and selected girth and fat measures are an important part of any fitness evaluation. They can be used to make a more accurate estimation of your desirable body proportions and body weight, and those subject to change are essential for setting your goals and evaluating your progress in any diet or exercise program.

Measurements should be taken stripped or in normal activity clothing with shoes, socks, and men's T-shirts removed. Whichever way they are taken, follow-up measurements should be taken in the same manner to insure validity of comparisons.

height, weight, and girth measurements

Height. By itself, of course, height is of no special significance. It is useful, however, as a factor in determining desirable body proportions and is necessary in conjunction with several other special evaluation items to be discussed. Normally, no changes in height would be expected as a result of an exercise program. But older persons and even younger ones who have lost flexibility in the spine or developed postural problems may be surprised to find that they add slightly to their stature, as did one sixty-two-year-old man in our adult fitness class who "grew" a half-inch following six months of intensive stretching exercises.

Weight. As we shall see, body weight per se is not nearly as important a measure of fitness as body fat, but it is nevertheless a useful guide. For the most accurate body weight starting point and later comparisons, it is best to weigh immediately after rising in the morning. Body weight will fluctuate throughout the day, depending on the effect of diet, activity, and environmental conditions and on water retention in the tissues, so sporadic weighing throughout the day will not give a true picture of body weight. Even consistent morning weighing will often show a fluctuation of from one to three pounds from these factors.

Girth Measurement. Girth measurements are made with a standard cloth tape measure as described below. Be sure to keep the tape level around the entire girth area.

Chest (normal). Erect standing position with arms at sides. Breathe normally. Place tape level around chest at nipple height.

Waist. Erect position. Breathe normally, without drawing in or protruding the stomach. Place tape band around waist at height of the umbilicus (belly button).

Hips. Erect position with feet together. Place tape around largest portion of hips.

Thigh (right). Erect position with feet approximately eighteen inches apart. Weight equally distributed. Place tape around largest part, usually in the crease just below the buttocks.

Calf (right). Same position as for thigh measurement. Place tape around largest part.

fat measurements

The amount of body fat a person has is considered to be a more reliable determinant of "overweight" or "underweight" than standard age, height, and weight tables. "Overweight" and "underweight" imply that we know what is normal weight for a particular individual, but actually we know very little about what is "normal." We would know what normal weight for an individual was only if tables were prepared from data taken from a population whose weight were considered normal for their particular age, height, and body type. However, most tables prescribing normal weight on the basis of age, height, and sex are based on "averages" taken from general population insurance statistics, and reflect only insurance risks. Though these tables may be valuable as a general guide for assessing a high risk premium, they do not accurately reflect weight normalcy in individuals with varying components of endomorphy, mesomorphy, and ectomorphy. A twenty-year-old six feet two inches extreme mesomorph may be at his best weight at 200 pounds and yet be considered overweight by standard tables. On the other hand, an ectomorph of the same age and height might be considered "normal" at 174 pounds and actually be 10 to 20 pounds "over-weight."

Attempts to list "desirable weights" by body frame still suffer from the same inadequacy of averaging within frame classifications. Furthermore, the terms "small frame," "medium frame," and "large frame" are poorly defined and are incomplete measures of endomorphy, mesomorphy, and ectomorphy. Another drawback of most standard weight tables is that they reflect slight weight increases with advancing age as being normal. Studies show, however, that this weight increase is caused primarily by an increase in body fat, which is not very desirable or healthy. Furthermore, since it takes five times as much fat as lean muscle mass to equal one pound, even a slight increase in body weight can mean significantly more fat. Some recent insurance tables have recognized this and now maintain maximum limits from about the age of thirty or thirty-five. This leads to the most serious limitation of standard weight tables, and that is the lack of close correlation between body weight and the amount of body fat, the major detriment to health and appearance.

Body weight and body fat relationship. According to Mayer, though relative body weight (under or above average) tends to be correlated with thickness of body fat, the correlation is not very close, "even in the wide range of from 30 percent underweight to 45 percent overweight."[3] Obesity, or excess fat, for young males is frequently defined as having 20 percent or more of body weight in fat tissues (27 percent for women). Though it is obvious, Mayer states, that obesity is present when overweight is markedly high, for moderate degrees of overweight, the relationship is not clear. He cites a study by Ancil Keys, a Minneapolis physiologist, which found that men considered twenty pounds underweight were actually in the upper-third of the distribution for fatness. Conversely, men 20 to 30 percent overweight were in the lower-third of fat distribution. Another study of fifty-one Air Force personnel compared by two standards, overweight (115 percent of standard body weight) and obesity (20 percent body fat), showed similar results.[4] Fifteen of fifty-one who were not 15 percent over the standard weight were nevertheless obese. Six cases which were considered overweight were found to have less than 20 percent body fat and were really not obese. Forty-one percent of the cases, then, would have been incorrectly classified on the basis of age-height-weight tables alone.

how fat are you?

The real concern in weight analysis, then, is determining how much body fat you have, since it is not only the sole criteria for obesity and best measure of risk to health, but it is also the component of body weight that can be changed most easily. Determination of body fat proportions can be made by certain direct and indirect measurements. Direct measurements, such as underwater weighing and chemical analysis, are the most accurate but are beyond practicality for us. Some simple,

---

3 Mayer, *Overweight*, p. 27.
4 Herbert A. de Vries, *Physiology of Exercise* (Dubuque, Iowa: William C. Brown Company, 1966), p. 227.

indirect methods can be used, however, with a fairly high degree of accuracy.

Mirror test. Looking at yourself unclothed in a mirror is often a more reliable guide for determining if you are too fat than scale weight. If you look fat, you are! Look particularly for loose and flabby skin in those areas where fat tends to show most: the mid-section, buttocks, and thighs—especially behind the upper part of the legs.

Pinch test. If you can't tell, the pinch test is quite reliable. It is estimated that in persons under fifty, at least 50 percent of the body fat is directly under the skin and, in general, should be between one-fourth to one-half inch in thickness, according to Mayer. By applying pressure with the thumb and forefinger, a fold of skin and the underlying fat can be pulled away from the bone and muscle tissue below. The total skinfold (skin and underlying fat) will be double thickness. Hence, depending on somatotype, any amount from one-half to one inch may be considered in the "normal" range: Ectos: one-half, Mesos: three-quarters, Endos: one inch. Any amounts above these should be considered excess. The good places to pinch are the abdomen (just parallel to a line from the nipple to the umbilicus), the front thigh (halfway between the knee and hip joints), and the biceps, triceps, and hip as described below.

Skinfold caliper measurements. The mirror and pinch tests are good practical estimates you can always make. Our most accurate practical guide has been skinfold caliper measurements taken at four sites: the hip, the triceps (back of the upper arm), the biceps (front of the upper arm), and the subscapular (just below the lower part of the shoulder blade). The total measurements at these four sites have been found to correlate very well with body fat percentages determined by underwater weighing. These correlations are shown in Table 1.

The cutoff point for obesity must still be arbitrary until further research can clearly determine what percentage of body fat is normal for various body types. Based on various standards and percent body fat and health risk statistics, the following cutoff points are suggested: Men —20 to 24 percent; women—27.5 to 32.5 percent. The figures are within standards set by the U.S. Air Force and below the 39 percent cutoff point for both men and women set by Mayer and based on a continuum of general population averages. For basic ectomorphs and medials, the cutoff point may very well be lower than those given, so for these persons body fat percentage estimates should be tempered by subjective judgment based on the mirror and pinch tests.

Making the measurements. To make these measurements you will need the assistance of another person. A standard fat caliper available at hospital supply stores would be most accurate, since caliper pressure

table 1
body fat content as indicated
by sum of skin folds at four sites*

| | | *—Fat as Percent of Body Weight—* | | |
Total Skin Fold (mm)	Men	Women	Boys	Girls
15	5.5	–	9.0	12.5
20	9.0	15.5	12.5	16.0
25	11.5	18.5	15.5	19.0
30	13.5	21.0	17.5	21.5
35	15.5	23.0	19.5	23.5
40	17.0	24.5	21.5	25.0
45	18.5	26.0	23.0	27.0
50	20.0	27.5	24.0	28.5
55	21.0	29.0	25.5	29.5
60	22.0	30.0	26.5	30.5
65	23.0	31.0	27.5	32.0
70	24.0	32.5	28.5	33.0
75	25.0	33.5	29.5	34.0
80	26.0	34.0		
85	26.5	35.0		
90	27.5	36.0		
95	28.0	36.5		

* Wayne D. Van Huss, Roy K. Niemeyer, Herbert W. Olson, and John A. Friedrich, *Physical Activity in Modern Living* (2nd ed.), © 1969. Reprinted by permission of Prentice-Hall, Inc., Englewood Cliffs, N.J.

can be set and the reading taken directly from the caliper dial. If available, so much the better. Otherwise, a small, inexpensive, hand-controlled pressure caliper purchased at a hardware or craft store can be used, with readings measured against the millimeter scale on an ordinary ruler—even though it will be slightly less accurate. If neither is readily available, a dime store divider bent so that it is curved at the ends can be used. By taking periodic measurements, even the less accurate measures in terms of computing body fat percentage can serve as a fairly precise measure of relative body fat gain or loss.

In taking the measurements, the thumb and forefinger of one hand (fingertips) are used to pinch and lift the skin and underlying fat tissue away from the muscle tissue below, using a moderate amount of pressure. The handle of the caliper is released to allow the full force of the caliper arm pressure and the reading is made to the nearest .5 millimeter. If a hand pressure operated caliper is used, it should be squeezed to the threshold of pain. Two applications are recommended for a stable reading. If skinfolds are extremely thick, readings should be made three

seconds after applying caliper pressure. To minimize the effect of variations from tissue dehydration, the measurements again are best taken in
the morning upon arising.

Hip. Pick up a skinfold halfway between the lower rib and hip bone
in a vertical line from the armpit.

Triceps. With the arm alongside the body, pick up a vertical skinfold
halfway between the elbow and armpit.

Biceps. Use same method as for triceps measurement.

Subscapular. Pick up a skinfold just below the shoulder blade, following the natural fold running parallel with the shoulder blade.

general endurance (working capacity)

General endurance or working capacity refers to your ability to work
for a long period of time before fatigue or loss of energy sets in and you
are forced to rest. Though somewhat interdependent, general endurance
is distinct from muscle endurance. General endurance is associated with
whole body activity, involving many muscle groups working simultaneously, such as in walking, running, or swimming. On the other hand,
muscle endurance refers to the ability of specific muscles or muscle
groups to work for a long period of time before becoming fatigued.
Local muscle endurance depends primarily on the overall strength of the
muscle and the condition of the small blood vessels within the muscle
itself, and is a valid and valuable measure of fitness to this extent.
General endurance, however, is considered to be the most important
single measure of a person's overall level of fitness because it strictly
reflects the condition of the lungs, heart, and entire cardiovascular system
—as well as the general condition of the muscles and the body's overall
health. An explanation will show why.

All activity requires energy. Essentially, energy for body activities
depends on chemical changes produced by combining oxygen and food
substances—primarily carbohydrates. Food may be considered the "fuel"
and oxygen the "flame" which ignites it. Since the body can store food,
but not oxygen, the amount of work it can perform depends primarily on
its ability to supply oxygen to meet the demands of the working tissues.
The stronger the muscles involved in breathing and the better the condition of the lungs, the more air can be brought in and the more oxygen
can be absorbed by the blood. The stronger the heart and better the
condition of the blood vessels, the more oxygen-carrying blood can be
brought to the working tissues. And the better the condition of the muscles, the more blood and oxygen they can absorb and the more efficiently
the oxygen is used.

Normally, when the oxygen supply is adequate to keep up with the energy demands of the body, the end products of the energy-making process are carbon dioxide and water which are easily carried off by the blood and removed through the lungs and other organs. This process of energy production is referred to as the *aerobic process* (with oxygen). For most people, it is the normal energy-making process in the resting state and in light exercise.

When there is an insufficient supply of oxygen for a particular type or level of activity, however, the energy-making process ends at an intermediate stage in which lactic acid is the end product. Energy produced in this manner is referred to as the *anaerobic process* (without oxygen— or an inadequate supply) and continues at the expense of creating what is called an *oxygen debt* which ultimately must be repaid. How soon the lactic acid buildup begins (anaerobic process) and how rapidly it builds depends on both the intensity of the activity (how much oxygen is demanded), the oxygen delivery capacity of the body, and the ability of the bloodflow to carry off the lactic acid for breakdown by other organs. Continued lactic acid buildup, however, ultimately upsets the chemical balance providing the immediate source of energy in the tissues. At this point, activity can no longer continue. It is at this stage that the point of exhaustion is reached, whether it be general exhaustion or local exhaustion of a particular muscle or muscle group. The body must then rest while the lungs and heart continue to bring in and pump supplies of oxygen to complete the lactic acid breakdown in the exhausted tissues —repay the oxygen debt. The huffing and puffing experienced after a strenuous exercise bout is simply paying back this oxygen debt.

For some persons, because of a weak oxygen delivery system (poorly conditioned lungs, heart, and blood vessels), the lactic acid accumulation starts early, builds up quickly, and the point of exhaustion is reached after only a small amount of general activity—perhaps little above that needed for continuance of vital processes and well below their normal daily activity requirements. Their period of recovery and repayment of their oxygen debt is also longer. Others reach the point of exhaustion only after prolonged periods of vigorous activity and are able to repay their debt quickly. The difference lies in the oxygen delivery capacity of their lungs, heart, and blood vessels, their capacity for removing the end products of the energy-making process, and the overall condition of their muscles—their general endurance or working capacity level.

measuring general endurance or working capacity

The best and most accurate measure of an individual's working capacity is the maximum amount of oxygen his body can consume. This can be measured precisely by laboratory techniques. Using a treadmill or bicycle ergometer to provide the work load, the individual is placed under a progressively increasing work load designed not to bring about local muscle fatigue before the maximum oxygen consumption level is reached. With the aid of a special mask, the expired air is collected in bags and periodic samples are analyzed to determine the amount of oxygen being consumed. When the calculations fail to show an increase in the amount being used, the point of maximum oxygen consumption is indicated. This point is usually associated with a close to maximum heart rate and a maximum rate and depth of breathing. The total amount of oxygen used, calculated from the total amount of the expired air collected, represents the person's maximum oxygen consumption and measure of his working capacity. Several other shorter, submaximal tests are often used to save time, with the results projected to estimate maximum oxygen consumption and maximum heart rate. In all these tests, the heart rate and behavior is carefully monitored by electronic devices and the test halted if any abnormalities appear.

The 12-minute run-walk. Obviously, such detailed laboratory measurements are too time consuming and impractical for mass use, and are reserved for research and special examinations. One of the most simple tests which correlates very well with laboratory measurements is the 12-minute run-walk field test developed by Dr. Kenneth Cooper and reported in his book, *The New Aerobics.*[5] Cooper reported a correlation of .90 with laboratory measurements. The test involves measuring the distance covered by running, or running and walking, for a total of twelve minutes. The distance covered is then used to rate a person in terms of his oxygen consumption or working capacity. Tables 2 and 3 show the ratings based on data collected on men from ages seventeen to fifty-two, and gives similar ratings for women.

Test procedure. First, a word of caution concerning the test. Heart rate and blood pressure cannot be monitored during the field test as it can in the laboratory. Persons over thirty, unless they have been exercising regularly for at least six weeks should postpone the test until they have had at least this much training. This is not only for heart safety, but also to avoid undue muscle soreness and fatigue following the test. No one should take the test without taking a medical examination and being cleared as normal. In any event, the test should be discontinued if extreme fatigue, shortness of breath, dizziness, or nausea are experienced.

For easiest distance measurements, it is best to take the test on a standard quarter-mile track or measured equivalent, divided and marked by eighths (every fifty-five yards). Otherwise, any measured area, similarly marked off, will do. Using a stopwatch or a watch with a sweep

5 Kenneth H. Cooper, *The New Aerobics* (New York: M. Evans & Company, Inc., 1970).

table 2*
12-minute test for men
distance in miles covered in 12 minutes

Under 30	30–39	40–49	50 plus	Rating
1.0	.95	.85	.80	Very Poor
1.0 −1.24	.95–1.14	.85–1.04	.80– .99	Poor
1.25–1.49	1.15–1.39	1.05–1.29	1.00–1.24	Fair
1.50–1.74	1.40–1.64	1.30–1.54	1.25–1.49	Good
1.75+	1.65+	1.55+	1.50+	Excellent

table 3*
12-minute test for women
distance in miles covered in 12 minutes

Under 30	30–39	40–49	50 plus	Rating
.95	.85	.75	.65	Very Poor
.95–1.14	.85–1.04	.75– .94	.65– .84	Poor
1.15–1.34	1.05–1.24	.95–1.14	.85–1.04	Fair
1.35–1.64	1.25–1.54	1.15–1.44	1.05–1.34	Good
1.65+	1.55+	1.45+	1.35+	Excellent

* From *The New Aerobics* by Kenneth H. Cooper, M.D.; © 1970 by Kenneth H. Cooper; p. 30. Reprinted by permission of the publisher, M. Evans and Company, New York, New York.

second hand to set the starting time, begin running at a slow to moderate pace. If you become short of breath, stop running and walk awhile until your breathing becomes easier again, then continue running. At the end of the twelve minutes, determine the number of completed laps and segments, calculate the total distance covered, and refer to the chart for your rating. If with another person, one can serve as the timer and recorder while the other is being tested, periodically calling out the elapsed time. The roles can then be reversed. In a group situation, the same type of pairings can be made, with the person in charge serving as timer. Periodic retesting during whatever training program you follow will help to measure progress. You should attempt to reach at least the "good" category.

Test modification. For persons under thirty, a modification devised by T. L. Doolittle and Rollin Bigbee[6] can be used. It was used in testing

6 T. L. Doolittle and Rollin Bigbee, "The 12-Minute Run-Walk: A Test of Cardiorespiratory Fitness for Adolescent Boys," *Research Quarterly*, 39, No. 3 (1968), 491–95.

ninth grade students and had a reported correlation of .94 with laboratory measurements. The modification involves walking, if necessary, no more than one-eighth of a lap at a time and calling out the remaining time at nine, ten, eleven, and eleven-and-a-half minutes. If able, the person being tested is urged to increase his pace during the last three minutes and put forth his best effort in the last minute. For younger persons and those in good condition, this increased intensity modification is good for more accurate results, but older persons or those in poor condition should not use it for safety reasons.

resting heart rate and heart recovery test

How high the heart rate is when the body is at rest, how high it climbs during various intensities of exercise, and how quickly it returns to its starting rate following different levels of exercise are good indications of your heart condition and working capacity.

Resting Heart Rate. Many factors will cause a variation in the Resting Heart Rate at any given time, such as level of prior activity, food intake, environmental conditions, smoking, emotions, illness, and even body position. It is lowest when lying, higher when sitting, and higher yet when standing. For practical purposes, the truest Resting Heart Rate can be determined by taking the pulse for one minute just after waking in the morning in the lying position. The pulse may be taken at the left wrist, in the cavity about two inches below the base of the thumb; at the neck, just to the side of the "Adam's apple" on either side; directly over the temple; or in the small cavity close to the ear at the junction of the upper curve. Some find it easy to take simply by placing the hand over the heart.

Just what a normal Resting Heart Rate should be is difficult to say. One study of healthy young men showed an "average" of 64 beats per minute, but a range of from 38 to 110.[7] The American Heart Association accepts as normal a range of 50 to 100. Studies have shown, however, that persons with resting heart rates of over 70 stand a much greater risk of heart attack than those with rates below 70, and that the risk increases and decreases proportionally as it is above or below 70. We also know that athletes and others in good physical condition, as measured by the working capacity tests described, have lower rates than those who perform poorly, and that the Resting Heart Rate can be reduced by regular exercise. A good minimum goal to set, then, is a

[7] Peter V. Karpovich and Wayne E. Sinning, *Physiology of Muscular Activity* (7th ed.) (Philadelphia: W. B. Saunders Company, 1971), p. 198.

Resting Heart Rate of 70 unless it becomes evident from the other tests that a higher rate is normal for you even when in good condition. Conversely, even if your Resting Heart Rate is below 70, it should not be considered satisfactory if the tests show otherwise. Your goal should be to lower and maintain it at the rate it reaches when in good condition.

Heart Recovery Rate. How high the heart rate rises during exercise and how quickly it returns to its starting rate are better indications of the heart's condition than the resting heart rate alone. Naturally, both will depend on the intensity of the exercise, as well as the condition of the individual. Strenuous exercise will cause a greater rise and the heart rate may not come back to normal for hours, while lighter exercise will bring a smaller rise and recovery within minutes. In either case, the better conditioned you are and become, the smaller the increase and the quicker the recovery.

A simple way of determining Heart Recovery Rate is through a step test. In this test, the subject steps up and down on a chair, bench, or platform of a certain height at a predetermined rate per minute and for a predetermined time. The postexercise heart rate is then determined by pulse counts at various time intervals and compared to certain standards or the starting rate. There are many standardized tests available, but they call for specific bench heights and many are quite fatiguing, especially on the legs. In addition, unless calculated by a complex formula, recovery norms are based on absolute values rather than in relation to starting rates. We prefer a test that is more relative, less fatiguing, and one that can be used at any time to estimate your present condition and measure your progress. It can also be used to determine whether or not you should postpone the 12-minute run-walk test.

Heart recovery test. Any standard size chair or bench, sixteen to twenty inches can be used for the test. Sit quietly for a few minutes and allow your pulse to settle, then

1. Stand and take a ten-second pulse count for your Starting Heart Rate.
2. Starting with either foot, step up and down on the bench or chair in a four-count movement at a rate of two complete up-down movements every five seconds for one minute (twenty-four per minute).
3. Immediately after completing the exercise, remain standing and take a quick ten-second count to estimate your Exercise Heart Rate (optional).
4. One minute after completing the exercise, take another ten-second standing pulse count to determine your Heart Recovery Rate.

Ratings. Recovery to Starting Rate or less = excellent; 1–2 beats away = good; 3–4 = fair; 5–6 = poor; 7+ = very poor.

Explanation. Notice we call the initial pulse count the Starting Rate rather than the Resting Rate, since it is not likely to be the same as your true resting rate, which is affected by the many variables mentioned. For the most accurate and consistent results for later comparison, however, an attempt should be made to keep all controllable factors constant (such as not eating for at least an hour before the test or not smoking just prior to it), but because the ratings are based relative to the Starting Rate, the variables for the most part are cancelled out. The significance of the Exercise Heart Rate at this point is just to get a general idea of how high your heart rate rises during this exercise, to use it as basis for future comparisons, and to provide practice for its use during the endurance training program that will be presented. Studies show that a quick ten-second count taken within fifteen seconds following any exercise will reflect within 90 to 97 percent accuracy the peak heart rate reached during the exercise period. Tests using bench heights of from 12 to 20 inches showed that one minute was sufficient time for the pulse to return to its starting rate following one minute of stepping at this rate. Our preliminary tests have shown a high correlation between the results of this step test, the 12-minute run-walk test, and percentile ratings of heart function measured electronically while subjects exercised on a bicycle ergometer, particularly at the lower and upper ratings. Significantly, not one smoker has made full recovery, regardless of his level of exercise activity.

Since the reliability of results on this step test depends on the accuracy of the pulse counts, some practice may be required before consistent results are found. Emotional anticipation may also affect the Starting Heart Rate in early trials, but with experience this factor can be eliminated.

In conjunction with the Heart Recovery Test, we also recommend recording a ten-second sitting Resting Heart Rate to serve as a practical Resting Heart Rate check at various times throughout the day and as a basis for future comparisons.

A person who scores in the "poor" or "very poor" category on the step test should postpone the 12-minute run-walk test until his recovery rating reaches the "fair" category.

vital air capacity test

The Vital Air Capacity Test is an excellent screening device for determining the specific condition of the lungs and degree of airway obstruction present—another key factor in working capacity. We use it as an additional check on those who score poorly on the Heart Recovery Test or 12-minute test, especially if they are heavy smokers or live or work in areas of high air pollution.

In the test, the subject inhales as deeply as he can and then exhales as forcefully as he can into a simple device called a Vitalor. The device records a tracing which shows the amount of air exhaled over a six-second period, up to a maximum of 4.6 liters. A more sophisticated instrument, called a spirometer, is used by specialists and can make maximum recordings over a longer period of time.

Two of the most important readings are the maximum amount of air that can be forcefully exhaled and the amount that is expelled in the first second. These are compared with predicted values based on age, sex, and height. Readings below predicted values, corrected for a 15 percent standard error of estimate, are strongly indicative of airway obstruction and lung disease, and would suggest further medical tests. In COPD patients, maximum forced vital capacity (FVC) is often significantly reduced. A below normal one-second reading ($FEV_{1.0}$) is considered the most valuable and reliable clue to the severity of airway obstruction. Of greater concern would be a drop in readings over a period of time, say six months or a year.

The test is recommended if you scored poorly on the Heart Recovery Test or 12-minute test and have any reason to suspect possible lung damage from smoking, occupational fumes, or high levels of residential air pollution. If unavailable elsewhere, it would be a good idea to ask your physician for the test. Persons with or recovering from a minor respiratory illness such as a cold or flu should wait until recovery before taking the test since results will be inaccurate.

cardiac risk index

An estimate of your chances of suffering a cardiovascular accident—heart attack or stroke—can be arrived at by completing the Cardiac Risk Index in Appendix A. The index includes most of the risk factors presented in Chapter 1 and was prepared by John L. Boyer, M.D. Dr. Boyer is a practicing physician specializing in internal medicine and is also the medical director of the San Diego State College Exercise Physiology Laboratory.

motor fitness

Motor fitness is one of the major components of physical fitness and includes such elements as muscular strength and endurance, power, flexibility, speed, agility, balance, and coordination. These qualities are not as directly vital as circulorespiratory fitness for general health, but play several important direct and indirect roles both in functional health and performance capacity.

As we have already mentioned, muscle weakness is a primary cause of poor posture and nagging low back pain. Lack of local muscle endurance, even with sufficient general endurance, can limit performance in certain types of work and play. Flexibility, speed, agility, balance, and coordination are factors in safe and efficient body movement which, in turn, can help prevent or avoid injury in certain situations.

Beyond a minimal level for functional health, safety, and emergencies, however, the extent to which you have or can develop these elements is important only in terms of the purpose for which they can be used—what do you want or need them for? Unlike circulorespiratory fitness which is so vital for the health of your heart, lungs, and for working capacity, a great amount of strength or muscular development serves no real purpose unless it is needed for a particular type of work or sports activity that you must or would like to do.

For example, take cardiovascular fitness and relaxation as possible major personal goals. Running, of course, would be the quickest way of improving cardiovascular function and releasing tension and the easiest, because it involves very little skill. But for many, running can become very monotonous, and may be less enjoyable than other activities which can provide the same benefit such as swimming, tennis, or handball. These activities, however, require skills based on various levels of motor fitness. Developing these qualities further can help you achieve your primary goals with far more enjoyment, likelihood of continuing, and added social and emotional benefits. It is in this way that high levels of motor fitness can indirectly contribute to your mental and physical well-being, but it is the activity for which it is used that is the direct contributing factor. Unrelated to some higher purpose, high development of motor fitness is not that important or a particularly major criteria of overall fitness.

Another reason why high levels of motor fitness in and by themselves cannot be set up as major criteria is that the extent to which they can be developed depends largely on inherited differences in body type and other body functions. Strength and speed, for example, are inherently limited, as are power and agility which are particularly dependent on them. Environmental background and previous experience also play a part. On the other hand, a high level of cardiovascular endurance is within the range of all—regardless of build or innate limitations on motor fitness.

evaluating motor fitness

Most elements of motor fitness are specific to the particular task being performed. This makes it difficult to make a generalization about a person's particular degree of strength, coordination, flexibility, or other motor fitness elements. To completely define, isolate, and measure all the elements in all their applications would be ideal, but it is an impractical and impossible task. The best that motor fitness tests can do is to measure some of the most important elements and give you a basic idea of where you stand in comparison to others.

Many good test batteries have been devised, and can be very useful to coaches, the military, police and fire departments, and others concerned with screening and checking candidates for positions requiring high levels of motor fitness. For our purposes, however, for the information that can be gained in comparison with the time and effort involved, they are not considered essential. The exercise programs in Chapter 5 are designed so that by following them a reasonable level of motor fitness for daily living will be assured. In applying the principles and methods of exercise to be discussed in Chapter 4, you can also adapt them to help you reach the highest level possible for your particular purpose.

activity competency

Your ability to engage in a wide variety of activities, games, and sports can make an important contribution to your overall fitness. Various activities tend to stress different components in varying degrees, and no one activity can usually satisfy all your needs for very long. An activity like bowling, for example, can be very emotionally and socially satisfying, but contributes only slightly to the vital cardiovascular and other motor fitness components of physical fitness. Vigorous team activities like football, basketball, or soccer may meet social, emotional, and almost all your physical needs for a time, but are usually limited to school and college years or shortly beyond them. In one way or another, a lack of competency in a wide variety of activities can limit your opportunities to achieve a balanced level of physical and total fitness.

As a guide in selecting activities you should attempt to learn, we have included in Appendix A an Activity Competency Questionnaire as part of your overall fitness evaluation.

your ability to relax

The ability to relax or relieve stress and tension is important to mental and physical health. The relaxation questionnaire in Appendix A is aimed at evaluating your degree of tension and your ability to gain relief from it. Specific methods and techniques for relieving stress and tension will be presented in Chapter 8.

general health and fitness checklist

Also included in Appendix A is a General Health and Fitness Checklist which summarizes some of the major factors that have a bearing on your level of overall fitness. Any negative responses would indicate the desirability of corrective measures.

the smoker's test

For those who have considered the advisability of quitting cigarettes, four Smoker's Self Tests are included in Appendix B. The tests are designed to determine what you know about cigarette smoking and how you feel about it—why you smoke and whether or not you really want to quit. The tests will also give you an idea of how difficult or easy you may find it to stop. Following the tests is a brief discussion aimed at providing you greater insight into the reasons for smoking and the difficulties involved in breaking the habit. Included is a description of various methods that have proven successful in breaking the smoking habit.

evaluation guides for drinkers

Most authorities agree that occasional and moderate drinking presents limited danger to most people. For many it can provide benefits in the form of pleasant relaxation and social and emotional satisfaction. The major damage from alcohol appears to stem from its habitual and excessive use, reaching devastating proportions in the alcoholic.

The decision to drink or not to drink is an individual choice, but it should be based on some basic understanding of the immediate effects of alcohol, the potential dangers involved, and a recognition of the symptoms that may lead to the most serious potential danger of all—alcoholism. For those concerned, Appendix B contains a brief discussion on some of the immediate effects of alcohol when various levels of concentration in the blood are reached, and on the causes, development, and warning signs of alcoholism.

**moving on to improving
your fitness level**

Before going into various exercise programs and other techniques for raising your fitness level, a discussion of the effects and benefits of exercise and the basic principles of exercise and training will first be covered. The next chapter will discuss the effects and benefits of exercise on the major systems of the body and in relation to some of the health problems and fitness components presented. The purpose of the chapter is to give you greater insight into how and why the benefits of exercise take place, a greater understanding of what has already been covered, and a background for understanding the exercise principles and programs, and the other fitness principles and techniques to be explained.

3

effects and benefits of exercise

Over 2,000 years ago, Hippocrates said, "Exercise strengthens while inactivity wastes." Today, it is a basic physiologic law that an organ or system improves with use and deteriorates with disuse.

The principle that the body develops with use and deteriorates with disuse has led many observers to see a cause and effect relationship between the decrease in the level of physical activity among our population and the parallel increase in the number of degenerative diseases and ailments. Though vigorous activity is no longer natural to our way of life, it is still a biological necessity. The best substitute for the lack of it in our normal pattern of living is a planned program of vigorous exercise faithfully carried out. Vouching for the beneficial effects of a planned program of exercise is a statement by the American Medical Association: "It begins to appear that exercise is the master conditioner for the healthy and the major therapy for the ill."

immediate, long-term, and residual effects of exercise

The physiological effects of exercise on the body can be classified into three types: immediate, long-term, and residual. The immediate effects are those changes that take place immediately as the body responds to the increased stress of exercise. An increased heart rate, faster and deeper breathing, and sweating are examples. These changes or adjustments take place regardless of a person's level of condition or how often he exercises. They are necessary adjustments to enable the body to handle the increased work load but are temporary. When the exercise ceases, the changes reverse themselves and the body returns to its normal state. The better conditioned the individual, the faster the return.

If exercise is continued on a regular basis, gradual changes take place in the body as a result of its repeated adjustments to the demands of exercise. These changes are of a more permanent nature, since they remain even after the body returns to its resting state. The gradual changes that take place in the body over a period of time as it adapts to the continued demands of exercise are called long-term effects. Since they remain more or less permanent as long as exercise is continued regularly, they are often referred to as chronic effects.

The changes that remain when exercise is discontinued for a long period of time are called residual effects. Although there is some evidence that animals trained early in life respond significantly better to exercise later in life than those who were not trained early, studies indicate that the chronic effects are transient unless exercise is continued on a regular basis. Changes involving the cardiovascular system and multiple motor fitness elements, such as agility and coordination, appear to reverse themselves at a faster rate than those involving single elements such as strength and flexibility.

effects on and benefits for
various systems of the body

It is important to keep in mind that the changes that take place in the various body systems in response to exercise do not take place independently of each other. Specific effects on one part in turn affect other parts. It is only for the purpose of clarity and understanding that a systematic breakdown is used in explaining these effects. Furthermore, since the effects of exercise could presumably be harmful as well as beneficial, an attempt is made to explain how the various physiological changes that take place can favorably influence the body—making it function more efficiently, better able to cope with the demands of modern living, and more resistant to degenerative diseases. Finally, it should be remembered that isolating a single factor such as exercise among the many factors involved in the maintenance of health is very difficult. There are many people who apparently enjoy good health and longevity without any more physical activity than their work allows. Nevertheless, there is sufficient clinical and experimental evidence to demonstrate the contributions of exercise to physical health.

cardiovascular system

Function. Nutrients, oxygen, and other materials needed by the body for its various metabolic activities are delivered by the blood. *Metabolic activity* refers to the chemical, changes that take place in the various cells of the body by which tissues are broken down, made, and repaired, other materials needed by the body are produced, and energy for all activity is provided. The blood also transports the waste products of cell metabolism to various organs for removal. In truth, the flow of blood has been called the "river of life." The function of the cardiovascular or circulatory system is to provide the necessary flow of blood to the various tissues of the body in response to their activity needs.

Components. The circulatory system consists of the heart, arteries, capillaries, and veins. The heart is a special four-chambered muscle about the size of a fist, which serves as both a storage tank and pump for moving blood through the system. Arteries are tubular vessels which carry blood away from the heart. The capillaries are tiny, single layer vessels located in the tissues themselves and are where the actual exchange of oxygen, nutrients, and waste products takes place. The capillaries then connect with a system of small, then larger, veins which return the blood to the heart for recirculation.

Blood flow. Actual blood flow through the system is accomplished by the alternate contraction and relaxation of the two lower chambers (ventricles) of the heart, which are governed by the involuntary nervous system. Each time they contract, blood is pumped from the lower chambers through two separate circuits leading back to the upper chambers (atriums). One circuit (the pulmonary circuit) carries deoxygenated blood from the lower right chamber to the lungs, where it gives up its carbon dioxide and other waste products, absorbs new oxygen, and is returned to the upper left chamber and waits for delivery to the working tissues. The other circuit (systematic circuit) carries fresh blood from the lower left chamber to all parts of the body and the deoxygenated blood with its waste gases back to the upper right chamber where it awaits recirculation through the lungs. When the heart relaxes, the lower chambers are refilled so that the blood can be moved with the next contraction. A condition in which the heart rhythm goes wildly out of control causing the lower chambers to vibrate rather than to contract is called *ventricular fibrillation* and is a common cause of sudden heart attack deaths. One of the dangers of smoking is that the effect of nicotine on certain nerve fibers in the heart lowers the threshold at which ventricular fibrillation takes place.

The familiar sound of the heart beat, "lub-dub," "lub-dub," "lub-dub," is actually the sound of the opening and closing of the one-way heart valves and the vibration of the heart as it alternately contracts and relaxes. The pulse also reflects this rhythmic heart action and is the result of the alternate raising and lowering of pressure against the artery walls as the heart pumps.

Blood pressure. In medical terms, the contraction phase of the heart action is called *systole* while the relaxation phase is referred to as *diastole.* Blood pressure is usually measured in millimeters of mercury and expressed in two figures. The first and higher figure represents pressure on the arteries during heart contraction and is called *systolic blood pressure.* The second and lower figure is the *diastolic blood pressure.* In young adults, blood pressure averages approximately 120/80. Both systolic and diastolic blood pressure generally increase with age.

In general, blood pressure is determined by two factors: cardiac output and resistance to blood flow through the arteries, capillaries, and veins. Cardiac output is the amount of blood pumped by the heart per minute, and is the product of *heart rate* and *stroke volume* (HR × SV). Stroke volume is the amount of blood pumped from the heart with each contraction. Blood pressure, then, will generally rise with either an increase in cardiac output (from a higher HR, SV, or both) or with an increase in resistance to blood flow. The increased resistance may be caused by either a decrease in the diameter or elasticity of the arteries, a poor exchange of blood between the capillaries and tissues (diffusion), a poor return of blood to the heart from the veins, or other factors.

With respect to blood return to the heart, as blood enters the capillaries, pressure drops considerably to allow a slower blood flow and a more efficient exchange of materials. Entering the veins, then, it is at a very low pressure. In the resting state, pressure is usually sufficient to provide an adequate return to the heart. During activity, however, movement of blood through the veins must have some assistance. The assistance is given by the skeletal muscles, particularly the muscles of the arms and legs. As these muscles contract, they squeeze the veins which helps to force the blood through one-way valves back to the heart. Thus, these muscles serve as auxiliary pumps. The action is called *peripheral heart action* and it is important to remember for several reasons. The heart can't pump what it doesn't have, so the importance of good peripheral heart action during vigorous activity, and especially immediately following it, can easily be appreciated. Dizziness from standing at attention for long periods of time or from rising quickly after sitting for a long time is a result of reduced peripheral heart action. For this reason, too, it is a good idea to break up long periods of sitting by moving the arms and legs or moving about for a short time.

Range of Systemic Arterial Pressure

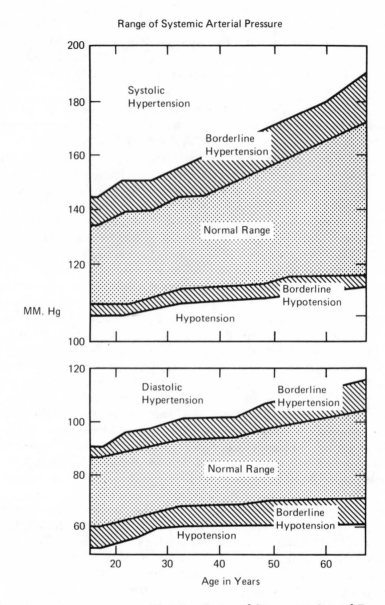

Fig. 6.   Range of Systematic Arterial Pressure
(From Rushmer, R. F.: *Cardiovascular Dynamics,* 3rd ed. Philadelphia, W. B.
Saunders Company, 1970.)

Coronary circulation. To complete the background information for
understanding the beneficial effects of exercise on the cardiovascular
system, some mention of coronary circulation should be made. Since the

heart is a muscle, it also requires a supply of blood for its activities. Blood flow through the heart muscle itself is called *coronary circulation.* Two large arteries on the right and left sides of the heart, with a branching network of smaller arteries and capillaries, provide the heart fibers with its supply of energy-producing oxygen.

If the supply of oxygen to the heart is inadequate for its work demands, its work must be restricted or heart failure will result. One of the dangers of high blood pressure is that it increases the amount of oxygen consumed by the heart during its work. To provide an adequate supply of blood, the heart must pump a greater total volume, usually by increasing its heart rate. But an increased heart rate only compounds the problem, since during heart contraction the coronary arteries are constricted. This means that these arteries are filled only during the relaxation phase. As the heart rate increases, there is a shortening of the relaxation phase and a decrease in coronary blood supply.

Immediate effects of exercise on the cardiovascular system. During exercise, there is an immediate increase in the heart rate and stroke volume, though the increase in stroke volume is very little until higher levels of activity are reached. The result is a greater volume of blood flow (cardiac output) to the working tissues, including the heart muscle itself. There is also a rise in systolic blood pressure so that more blood may be forced through the capillaries. The blood vessels (arteries, capillaries, and veins) enlarge and additional capillaries open up to allow more blood into the tissues. Circulation of oxygen-carrying red corpuscles increases, as does the level of sugar, cholesterol, and other fatty acids in the blood, because of the greater need for energy-producing materials. The level of carbon dioxide and lactic acid (waste product of energy metabolism) rises. Finally, the flow of blood out of the capillaries, into the veins, and back to the heart is increased by peripheral heart action of the skeletal muscles, enabling the heart to pump more blood back to the working tissues.

Long-term effects on the heart. One of the most beneficial long-term effects is an increase in the size and strength of the heart muscle and a higher stroke volume at all levels of activity. Because of the increased stroke volume, the heart does not have to pump as fast in order to supply the body with a given amount of blood for a particular level of activity. The slower heart rate at all levels of activity, resting and during exercise or heavy work, allows the heart more rest between contractions. It also provides for a greater blood flow through the coronary arteries to the heart muscle itself, assuring it of an adequate oxygen supply. In the well-conditioned individual, the heart rate also responds more quickly to various work loads and recovers faster to its normal resting rate. Because of its ability to supply an adequate amount of blood at a lower heart rate, the heart is able to tolerate greater work loads before reaching its maximum rate. For the average person, this can provide a reserve capacity to cope with sudden physical or emotional stresses which accelerate the heart rate. In a well-trained athlete, a substantial amount of vigorous activity over a relatively long period of time is required to raise the heart rate to its maximum.

Other long-term effects on the heart include an increase in the size and elasticity of the coronary arteries and an increase in the size and number of functioning capillaries. Latent ones are opened and new ones are formed. The increased size of the coronary arteries provides a greater blood flow under less resistance, and can help prevent or delay the closing of these arteries caused by the accumulation of fatty deposits on the linings. The greater size and number of capillaries provides the heart with more nourishment (which accounts for its increased size and strength) and with a better distribution of oxygen for its energy needs, in turn increasing efficiency. They also allow for easier removal of waste products and delay of fatigue which waste product buildup can cause.

By increasing the capillary beds, regular exercise also helps to develop collateral circulation in the heart. *Collateral circulation* refers to connections between arteries and veins of the same diameter and between larger and smaller vessels. It is important in continuing or reestablishing circulation when a main vessel is closed off for any reason. Its significance is best illustrated by an autopsy report on astronaut Jim White, who was burned to death in a tragic spacecraft cabin explosion during a test at Cape Kennedy in 1967. The autopsy showed a complete closing of one major coronary artery from atherosclerosis. Yet, because of a large buildup of collateral circulation, White had been able to function at a high level of physical efficiency. The autopsy findings also indicate the complex factors involved in determining the causes and prevention of atherosclerosis.

Long-term effects on the vascular system. As in the heart, there is an increase in the size and elasticity of the arteries throughout the body and an increase in the size and number of capillaries in all the various tissues. Collateral circulation between all major arteries is increased and there is better diffusion between the capillaries and the working tissues. The improved blood flow takes an additional load off the heart. The enlargement and increase in the number of blood vessels helps decrease resistance to blood flow and to lower an abnormally high blood pressure, both at rest and during exercise. The inherent dangers to the heart and the chances of vascular explosions anywhere in the system are thereby decreased. The effect also helps blood pressure to return to normal faster following an increase from any cause. By increasing the size and helping to maintain elasticity of the large arteries, hardening of the arteries can also be prevented or delayed.

Long-term effects on the blood. Regular exercise results in an increase in the number of red corpuscles circulating in the blood. The improved oxygen-carrying and waste removal capacity further increases work load capacity. Two other beneficial long-term effects relate to the formation of blood clots and the level of fatty acids in the blood. There is evidence that regular exercise affects the clotting time of blood in two ways, both beneficial to the body. First, there is a quickening of the coagulation of blood when it leaves the vascular system, such as from a cut. The aid in lessening blood loss from such cuts is obvious. There is also an increase in the activity of the blood plasma which breaks up blood clots forming within the system. This lessens the chance of a blood clot forming in the heart, brain, or other vital organs.

Numerous studies involving both animals and humans have shown that residual levels of cholesterol and other fatty lipids in the blood can be significantly reduced as a result of vigorous exercise. Dr. H. J. Montoye concluded in a report summarizing his findings that "in various populations, blood cholesterol, body fatness, percentage of calories from fat, and total fat intake all appear to vary together, and are inversely related to physical activity." Cardiologist and exercise physiologist, Dr. Lawrence Lamb, reported that examination of athletes competing in the grueling pentathalon event revealed a remarkably low blood cholesterol level, even though they normally consumed a high cholesterol diet of up to 6,000 calories a day.

The process of accumulation of fatty deposits in the arteries remains somewhat of a mystery and target for further research. Exercise, diet, and stress all seem to have an effect on blood cholesterol and triglyceride levels and the clear cut and relative effects of each are difficult to determine. Exercise and a low fat diet seem to lower these levels, while stress appears to raise them. The research would indicate, however, that exercise can have a beneficial effect.

The immediate and long-term effects of regular exercise on the cardiovascular system, as outlined, demonstrate why the incidence of cardiovascular disease has consistently been found to be lower in physically active people than in those who lead more sedentary lives. The effects should also help to explain why an increasing number of doctors are prescribing exercise for patients showing signs of degenerative cardiovascular disease and even for patients who are recovering from certain types of heart attacks. Lieutenant Commander Benjamin F. Gibbs, head of cardiovascular surgery at San Diego's Naval Hospital, has said that, based on his experience, he believes that three out of four patients suffering from cardiovascular disease can avoid surgery by following three recommendations: exercise regularly, reduce body weight, and stop smoking.

the respiratory system

Function and components. It has been pointed out that the energy required for all of the body's activities comes from the combining effect of oxygen, nutrients, and other materials. While nutrients and other materials can be stored, oxygen must be continually supplied. Since the body cells are not in direct contact with the external environment, they must rely on the respiratory and cardiovascular systems to deliver their oxygen needs and dispose of the waste gases of energy metabolism. The ability of both systems to deliver oxygen, then, is the most critical limiting factor in the nature and amount of activity the body can perform.

The entire respiratory process involves three separate but interrelated functions: (1) the exchange of gases between the environment and the lungs (*external respiration*); (2) the transport of gases between the lungs and the various tissues (a function of the cardiovascular system); and (3) the actual exchange of gases between the blood and tissue fluids (*internal respiration*). Our concern is with the effects and benefits of exercise on the external respiration process, which includes the lungs and the muscles involved in breathing.

Since the lungs have no muscles of their own, they are completely dependent on the muscles of the rib cage and the diaphram for their expansion and contraction. When these muscles contract, the ribs move outward, the chest cavity is enlarged, and the pressure inside it decreases. The greater pressure of the outside air then forces air into the lungs. When the muscles relax, the opposite takes place—the chest cavity decreases, the pressure becomes greater inside than outside, and the air is pushed out. The contraction and relaxation of these muscles is controlled by both the voluntary and involuntary nervous system.

Immediate effects of exercise. An immediate effect of exercise, of course, is an increase in the rate and depth of breathing in response to the body's need for more oxygen and the removal of increased levels of carbon dioxide. In addition there is an additional flow of blood to the tissues of the respiratory muscles and through the lungs themselves. The increased depth of breathing forces the use of more alveoli, the tiny air sacs in the lungs which connect to the capillaries where the actual exchange of gases takes place.

Long-term effects. Because of the increased blood supply, the muscles of respiration increase in size, strength, and endurance. Their increased strength and tone increases the size of the thoracic cage (chest cavity) at all levels of activity, allowing more air to enter the lungs with each breath. And they become more efficient, requiring less oxygen to perform a given amount of work. There is also an increase in the size and number of capillaries in the lungs, allowing more oxygen to be absorbed and more carbon dioxide removed. In addition, a greater number of alveoli are brought into operation, bringing about an even greater exchange. The increased number of both increases the efficiency of breathing. Less air is needed because it is being used more effectively and being absorbed into the bloodstream faster. The combined effect of deeper and more efficient breathing results in a slower rate of breathing in the resting state and during submaximal levels of activity—with a consequent saving of expended energy. The respiration rate returns to normal more quickly following exercise. The overall effect, in terms of physical performance, is the ability to perform more work and to work for longer periods of time without fatigue.

the muscular system

Function and components. All body movements depend on the alternate contraction and relaxation of muscle tissue for their driving force. Muscle tissue has the unique quality of elasticity, which accounts for its ability to contract, relax, and stretch. There are three major types of muscles: (1) *striated muscle,* composed of long, cylindrical fibers; (2) *cardiac muscle,* which consists of a criss-crossing network of striated fibers; and (3) *smooth muscle,* found in the blood vessels and hollow walls of the internal organs. Cardiac muscle is found only in the heart. The skeletal muscles of the body which provide the force for moving the skeletal bones and stabilizing body parts are made up of striated muscles. These muscles also provide smaller movements, such as movement of the eyes, face, and throat. The smooth muscles are controlled by the involuntary nervous system while the skeletal and certain other smaller striated muscles can be moved at will and are called voluntary muscles.

Skeletal muscles. The skeletal muscles are made up of bunches of fibers held tightly together by connective tissue. The type and number of fibers and bunches in a particular muscle depend on the body part it is to move and the amount of force required to move it. For movement of the larger body parts, one end of the muscle is anchored by connective tissue, usually a tendon, to a bone which it cannot move.

This attachment is called the *origin* of the muscle. The other end is attached to a bone which the muscle is intended to move when contracted. It is called the *insertion*. When the muscle contracts, it moves the bone toward its origin. For example, the bicep muscle has its origin at the shoulder bone and inserts at the bone just below the elbow on the thumb side of the front forearm. When it contracts, the lower arm is moved toward the shoulder. To return the arm, the tricep muscle, with a similar origin and insertion on the back side of the shoulder and forearm, contracts. All muscles moving the body parts work in similar pairs—there are about seventy-five pairs. As one contracts to provide the force for movement, the other (called the *antagonist*) stretches to provide a controlled, smooth movement. Other muscles may assist particular movements by providing additional power, fixating movable body parts, or controlling the force of movement.

Nerve control of muscle action. Muscle fibers are triggered to action, voluntarily or involuntarily, by individual connections to nerve fibers. The ratio of nerve to muscle fibers varies with the degree of precision required. In the eye, for example, a one-to-one ratio may be found. In the larger skeletal muscles, one nerve fiber, together with its branches, can activate up to 150 fibers. The nerve fiber with its branches and the muscle fibers controlled by them form what is called a *complete motor unit.* When a particular motor unit is stimulated, all the muscle fibers in the unit are activated. The larger skeletal muscles have many motor units.

The effects and benefits of exercise on the heart and smooth muscles were explained previously. Some of the effects on the skeletal muscles were hinted at. What concerns us now are the effects on the skeletal muscles, which are so essential for movement and general posture, and which give shape and contour to the body.

Immediate effects. When muscles are exercised, there is an increased blood flow through the capillaries into the muscle tissue. Depending on the type and severity of the exercise, this causes a temporary increase in the size of the muscle. Additional motor units, normally not activated, are brought into operation. Because of the increased metabolic activity in the muscle tissue, there is an increased production of heat which is reflected in a rise of body temperature. The increased temperature decreases the viscosity of the muscle; that is, it makes the molecules of the muscle less resistant to rearrangement. The muscle becomes more pliable and is able to contract and relax more easily and at a faster rate. As previously explained, peripheral heart action of the muscles is also increased during exercise. Vigorous exercise has also been shown to reduce levels of neuromuscular tension as measured by the amount of electrical activity in the muscles before and immediately after exercise.

Long-term effects. Depending on the type and amount, regular exercise increases the size, strength, and endurance of the muscles. In activities such as walking, running, cycling, etc., the major effect is on the flow of blood through the tissues. More efficient delivery of oxygen and removal of waste products allows the muscles to work longer without fatigue. The improved nourishment helps to keep the muscles in good tone (tautness). In power-type activities, such as weight training and heavy calisthenics, a thickening and increase in the size of both the muscle and connective tissue (*hypertrophy*) is more prominent. With this comes an increase in strength. Contributing to the increase in strength is the activation and development of additional motor units in the muscle. The increased strength and number of motor units also improves the efficiency and endurance of the muscles in another way. Motor units work in relays. As strength increases, fewer are needed to perform submaximal efforts. This allows the other units to rest. When the working units become fatigued, the additional units are brought into operation, allowing the tired units to rest. The result is a significant increase in the endurance of the muscle.

skeletal system

There are about 206 bones of varying shapes and sizes in the normal adult. Besides supporting and giving general shape to the body mass, the bones also serve as a protection for the softer and more vulnerable organs and systems such as the brain, lungs, and spinal cord. They also provide anchors for muscles and leverage for movement. The inner parts of certain bones (*marrow*) also serve the important function of producing blood cells and storing calcium. Vigorous exercise, by improving the supply of blood to the bones helps them to grow in size, thickness, and strength and to maintain their vital inner activity.

the digestive system

Function and components. Before food nutrients can be delivered to the various tissues, the food must be broken down until it is in a form that can be absorbed by the blood. The mechanical and chemical process by which this change takes place is called *digestion*. The mouth, esophagus (passageway from the mouth to the stomach), stomach, and small and large intestines form a continuous canal through which the food passes during the digestive process. Wave-like contractions of the smooth muscles assist the movement along this path. The liver and pancreas, two large glands, secrete substances which are essential for the chemical breakdown of the food.

Effects of exercise. Vigorous exercise approximately one hour before or after the intake of food interferes with the digestive process. Blood supply is diverted from the digestive system to the active muscles. The secretion of digestive juices is inhibited. In addition, the wave-like contraction of the smooth muscles along the alimentary canal is inhibited. The most notable effects are possible cramps and a delay in the emptying time of the stomach. Mild exercise after a meal, however, has been found to speed up the emptying time of the stomach and have little or no effect on the secretion of juices. Regular exercise, by strengthening the smooth muscles along the digestive pathway, can help to regulate bowel movement and prevent or relieve constipation.

endocrine (glandular) system

The endocrine system is related to the hormone secretions of certain glands. The word *hormone* comes from the Greek word meaning "to excite." The hormones are substances which excite cells into activity. When stimulated by the nervous system, the cells of the gland secrete raw material from the blood and manufacture the hormone. The hormone is then transported by the blood to specific cells which are then activated.

Effects of exercise. Vigorous exercise stimulates the production of certain hormones which have significant effects on the body. Stimulation of the adrenal glands produces two substances called *adrenalin* and *noradrenalin*. These adrenalins release caloric energy to specific cells and allow them to operate at higher levels of activity. In addition, they are partly responsible for the increased heart rate, blood pressure, and blood sugar needed for greater activity. An increase in the production of adrenalins can also be triggered by mental or emotional stress such as fear, anger, anxiety, or other emotional excitement. When this happens, the responses in the body are the same: increased release· of energy and rise in blood sugar, blood pressure, heart rate, etc. This response is often called the "fight or flight" response. The body interprets the feelings as a "threat" and the response serves to prepare it for appropriate action. In the absence of body action, however, the response is detrimental. The body is keyed up and poised for action. Unreleased energy is built up and muscles tighten. There is an unnecessary waste of energy and an unnecessary taxing of the heart and certain specific organs of the body.

Hans Selye, a Canadian endocrinologist, has theorized that every individual inherits a certain amount of what he calls "adaptation energy" in response to stress of all forms—physical or emotional. When this adaptation energy runs out, the body reaches the state of exhaustion; illness and, ultimately, death follows. He further points out that humans die primarily because of weakness or uneven wear of one organ or system, as a result of continual adaptation to stress.

Exercise can prove valuable in lessening the detrimental effects of adrenalins produced by emotional stress in several ways. According to Seyle, when stress is disproportionately greater on one organ or system, exercise may help by distributing the stress over a wider area. Vigorous exercise has also been found to reduce the physiologic effects of emotional stress. In an experiment conducted on college women basketball players, stress level was measured after a game and also after preparing for a game which did not materialize. It was found to be lower after the game that was actually played, indicating that the physiological effects of emotional stress can be "worked off."[1] Experiments have also shown a reduction in muscular tension, often a reflection of emotional stress, following exercise. Another important value of exercise in combating the effects of emotionally produced adrenalins is the strengthening of the body, particularly the heart, so that it can better handle the increased work loads. Because it is stronger, the heart may be able to resist attacks of emotional stress which might otherwise bring about heart failure and death.

Vigorous exercise also stimulates the thyroid gland which produces a substance called *thyroxin*. Thyroxin has a more general effect on the body than the more specific-acting adrenalins. It releases caloric energy and helps to activate cells in all parts of the body. The increased metabolic activity lasts for hours, even after exercise is stopped. The calorie expenditure resulting from this general rise, combined with the expenditure produced by the adrenalins, is an important factor in weight loss.

Over a period of time, by its affect on thyroid gland activity, exercise can help to normalize an individual's basal metabolic rate. A low rate may be raised and a high one lowered. The *basal metabolic rate* represents the minimum amount of energy used by the body in its resting state. Persons with a low basal metabolism tend to be lethargic. By raising the metabolic rate, the lethargy can be reduced. Tense people tend to have a high metabolic rate. Lowering it may help to reduce tension in these people.

Exercise also stimulates the secretion of *cortisone* from the adrenal cortex gland. Cortisone acts directly on the tissue of the brain and, therefore, can help to stimulate better mental activity.

The beneficial stimulating effect of exercise on the glands is summarized by this quotation from the *Journal of the American Medical*

---

1 Herbert A. de Vries, *Physiology of Exercise* (Dubuque, Iowa: William C. Brown Company, 1966), pp. 165–68.

Association: "The greatest value of exercise lies in its stimulating effect on endocrine activity, perhaps the thyroid, in particular, and in overcoming the tendency to sleep and snooze too much, a counterpart of obesity." The fatigue of old age is often the result of a drop in endocrine activity.

the nervous system

The nervous system initiates, coordinates, and directs all the activities of the body. It consists of the brain, spinal cord, and a network of nerves which branch out to all parts of the body. Nerve impulses travel back and forth from the brain to the various parts of the body with lightning speed. The *somatic nervous system* controls the operation of the skeletal muscles and is more or less under conscious control. The internal functions of the body, such as the operation of the digestive system, the heart rate, and blood pressure are controlled by the *autonomic nervous system*, over which we have no conscious control. The *sympathetic* division of the nervous system has to do with the "fight or flight" type of adjustments to dangerous situations and various types of stress as previously discussed (increased heart rate, etc.). The *parasympathetic* division controls the more continuous type of activities or adjustments such as the digestive system, and also serves to counteract and balance the reactions caused by the "fight or flight" stimulation of the sympathetic nerves.

The sympathetic nervous system responds to vigorous exercise, as well as to smoking, tension, and emotional stress. Exercise and physical activity, however, have a beneficial effect on certain parts of the balancing parasympathetic nervous system which the other forms of stimulation do not have. It has been found to particularly benefit the vagus nerve, which serves to inhibit or slow the heart rate. This means that regular vigorous exercise can help to balance out the detrimental effects of sympathetic nerve stimulation from smoking, tension, and emotional stress. With respect to the heart, the increased heart rate caused by these stimuli will be less because of the improved inhibiting effect of the vagus nerve.

Exercise has several beneficial effects on the somatic nervous system controlling muscle movement. One is an improvement in muscular coordination caused by the development of more efficient neuromuscular patterns or pathways. When you first learn a motor skill, for example, you have to think about every movement. The brain directs every movement of the muscles by sending specific nerve impulses. It receives feedback information via other nerves as to what the muscles are doing, rapidly sorts the data, and sends further impulses directing the muscles to the desired movement. As the movement is repeated over and over again, specific nerve patterns are established between the brain and the muscles for the specific movement. This information is stored in a certain part of the brain and future movements not only become almost automatic in response to the proper stimuli, but also more efficient because unnecessary movements are eliminated.

Reaction time has also been shown to increase as a result of repeated motor patterns. *Reaction time* is the time interval between the presentation of a stimulus and the response. Improvement in reaction time, however, appears to be specific. For example, improvement in a reaction involving the arms will not necessarily bring about an improvement in a movement involving the legs. Similarly, a person may be quick in a particular type of arm movement, but slow in another type of movement with the arms. Based on studies involving college age groups, reaction time for particular movements appears to be slower in women than in men. Both younger and older than college age groups were also found to have slower reaction patterns. The differences may well be caused by differences in activity patterns rather than by inherent differences.

the excretory system

The excretory system removes waste products from the body. It includes the lungs, the kidneys, the digestive system, and the skin. Most of the effects of exercise on the system are immediate.

The increase in body temperature causes an increase in sweating, mainly for its cooling effect on the body from vaporization in the atmosphere. In the absence of atmospheric conditions which allow evaporation, such as high humidity, the cooling effect is lost. Excessive loss of salt from sweating can bring about heat cramps and exhaustion.

Reduced blood flow to the kidneys lessens the production of urine. Severe exercise which builds up an excessive amount of lactic acid in the blood puts an extra burden on the kidneys. Normally, however, if the oxygen supply is sufficient, it is broken down into carbon dioxide and water and is easily removed through the lungs, kidneys, and skin.

general benefits

The many health and functional benefits of exercise so long proclaimed by physical educators and medical personnel are the result of the effects of exercise on the various systems of the body as briefly explained in the previous sections. The explanation, though necessarily an oversimplification of very complex matters, should help you to better understand why and how these benefits take place. Hopefully you will gain a greater appreciation for the marvelous organism that the human body is, be stimulated to greater inquisitiveness concerning its operation, and be more fully aware of the need to take care of it.

A summary of the health and functional benefits from the effects of exercise on the body include:

1. The prevention, delay, or greater ability to withstand and recover from cardiovascular accidents and other degenerative diseases
2. Control of body weight and proportions
3. Relief of tension
4. Improved posture and appearance
5. Prevention of low back pain
6. Delay of the aging process
7. Increased ability to meet emergencies requiring physical strength and endurance
8. Improved neuromuscular skill and physical performance
9. Improved functioning of the internal organs
10. Better digestion and bowel movement
11. Stimulation of mental activity
12. Higher levels of fatigue and greater physical work capacity
13. Increased general feeling of health and well-being

The specific details of many of these benefits will be presented in later chapters. Some have already been explained or touched upon. A few warrant further discussion at this point.

exercise and weight control

There are many misconceptions about the value of exercise in controlling body weight and maintaining proper proportions of fat and muscle tissue. Many people ridicule attempts to lose weight through exercise by giving exaggerated figures regarding the amount of exercise needed to lose just one pound. They love to give examples of having to walk at four miles per hour for ten hours or play handball for five hours just to lose one pound and then justify their lack of effort by the obvious impracticality of performing these feats. These people are also quick to point out that after exercising any benefit in calorie expenditure is lost by an increased appetite and calorie intake. Both of these arguments are faulty.

Laboratory studies on animals, verified by observations of humans, indicate that appetite increases and decreases proportionately to the level of activity only within a certain range of activity level, described by

Mayer as the "normal range of activity."[2] The minimal level of this normal range appears to be the equivalent of at least one hour of vigorous physical activity a day. Both animals and humans who fall below this minimal level fail to show a corresponding loss in appetite and food intake. In fact, it has been found that sedentary persons take in more calories than those who are mildly active. The excess calories are stored as body fat. The upper level of the normal range of activity appears to be the stage at which exhaustion sets in. When exercise is carried to this point, even in a well-conditioned person with normal body weight and proportions, appetite falls off, food intake decreases, and weight loss occurs. Sedentary persons, unaccustomed to vigorous activity, also tend to show a decrease in appetite and food intake in the early stages of an exercise program and consequent weight loss. Only within the normal range does weight stabilize and appetite and food intake respond proportionately to the level of activity. The stabilization of body weight in well-trained athletes, even though they may consume up to 6,000 calories a day, is further proof of the effect of exercise on weight control.

As for the argument concerning the amount of exercise required to lose just one pound, the fallacy is a failure to take a long-range view. It has been shown by Mayer that just one-half hour a day of vigorous exercise such as running or playing handball or the the equivalent represents the calorie equivalent of from sixteen to twenty-six pounds in a year. One hour a day of such activity, if diet remained constant, would amount to a weight loss from exercise alone of up to one pound a week. For the sedentary person, combined with the decrease in appetite accompanying the start of an exercise program and consequent lowering of the excess calories he normally consumes, this could mean a loss of up to a pound-and-a-half to two pounds a week until his body and appetite adjusts to the new level of activity. Furthermore, developing a habit of from a half to an hour a day of vigorous activity can help to prevent the gradual accumulation of body fat and weight which generally results from a failure of the appetite to decrease as the activity level falls. The long-term or gradual effect of exercise on weight control is an essential concept to keep in mind. An elaboration of these points and a more detailed look at the essentials of weight control is presented in Chapter 7.

exercise and aging

Certain measures of physical performance show a gradual decline with age, usually after age thirty. The rate of decline, however, varies both

[2] Jean Mayer, *Overweight* (Englewood Cliffs, N.J.: Prentice-Hall, Inc., 1968), p. 73.

with different individuals and with different functions, and it is hard to determine the relative effects of heredity, aging per se, increased sedentary habits, or unrecognized disease processes (such as athero- sclerosis). Certainly some decline in physical performance can be ex- pected with advanced age, but there is evidence that this decline can be slowed down by a sensible program of exercise. Furthermore, the decline is not enough to seriously restrict reasonable physical activity as one grows older. In fact, experience shows that persons in good health, particularly those who exercise regularly, require few restrictions.

An outstanding example of what is possible even at an advanced age is that of Clarence De Mar, a famous marathon runner who made it a lifetime habit to run twelve miles a day. Up to age sixty-five, he engaged in twenty-six-mile marathon races. An autopsy performed after his death at age seventy (from cancer) showed an unusually well-developed heart with coronary arteries up to three times the normal size. Larry Lewis, a San Francisco waiter, is another outstanding example. Making daily running a habit since the age of nine, he celebrated his 102nd birthday in 1970 by running 6.7 miles in thirty-seven minutes! At 101, he was clocked in 17.8 for a 100-yard dash on the University of California, Berkeley track. Many other less spectacular examples are often seen among sports personalities, physical educators, doctors, and people in other areas of life. The common factor among all of these people is the practice of a continuous program of physical activity—not a sporadic one.

The major benefit of such a continuous program is the slower rate of decline in the circulatory, respiratory, muscular, and other systems of the body. Well-controlled studies have demonstrated significant decreases in vital functions as a result of extended bed rest (from two to six weeks), in well-conditioned athletes. Among these are a decrease in stroke volume, an increase in heart rate, lower cardiac output, lower oxygen consumption level, and a loss of muscular size, strength, and tone. It follows that a sedentary life pattern, halfway between bed rest and a reasonable level of activity, will bring about a faster decline than a life that includes regular vigorous activity, and that such an activity program will prevent or delay the onset of degenerative diseases usually associated with old age. On a more positive note is the lengthening of the ability to engage in an active and enjoyable working and recreational life. Many older persons who have continued a regular activity program show more energy, vitality, and higher levels of physical performance than many of a much younger age. They also retain a trim and youthful appearance and almost certainly continue to enjoy life to its fullest.

Aging and body weight. As mentioned in the chapter on fitness evalu- ation, most standard weight tables show an increase in body weight with age as being normal, but studies show that this increase is caused primarily by an increase in body fat. One study of body weight and proportion changes from age twenty to fifty-five showed a gain in body fat of over 300 percent compared to an actual decrease in fat-free weight (muscle tissue). From what we know about the effect of both exercise and diet on body weight and proportions, it is difficult to attribute these changes exclusively to the aging process. Well-documented studies showing a decrease in basal metabolic rate with age, often cited as a reason for weight gain, may also be open to question as to the possible effects of exercise. Tests of metabolic rate showing this decrease with age are normally taken on the basis of total body mass: fat and fat-free tissue. Tests taken on the basis of active muscle tissue only, however, showed no change in the basal metabolic rate. The effect of exercise on both the loss of fat free tissue and the lower metabolic rate in old age is an interesting topic for further research.

ability to meet emergencies

The physical performance level needed by each person varies with the nature of his normal activities. A professional football player obviously needs a higher level of physical condition than an office worker. In life, however, many situations arise which call for unusual efforts. If the body is not prepared to meet these unforeseen demands, serious disability and even death may result. Examples of such uncommon, but possible situations are:

1. Encountering natural catastrophies (earthquakes, flood, fire)
2. Having a serious accident or injury requiring emergency treatment
3. Undergoing a serious operation
4. Falling into the water fully clothed or having to rescue someone who has fallen in
5. Dodging an oncoming vehicle or having to prevent someone from being struck by one
6. Having car failure in a remote and desolate spot
7. Needing to perform a short sprint at maximum speed to catch a public vehicle or to pursue someone

Many more examples can be added, but the point is clear. In situations of this type, all the body's resources and reserve capacities are called upon. The person whose body has been accustomed to the physical stress of exercise and who has developed high reserve capacities is in a much better and safer position to handle these situations than one who has not trained his body by exercise.

The exercise programs to be presented will help you realize the effects and benefits discussed in this chapter. But beneficial as they may be, exercise programs can be ineffective and even dangerous unless prepared and followed under the guidance of sound physiologic principles. These principles, along with helpful hints on how to make an exercise program safe, effective, and interesting enough to be of lasting benefit are covered in the next chapter.

4

exercise principles, techniques, and precautions

Principles are broad guidelines for action. The purpose of this chapter is to present some of the essential principles, techniques, and precautions that should be followed in any exercise program designed to improve or maintain a particular level of physical condition. This knowledge will help you not only to safely achieve more effective results, but also to better understand and evaluate the programs to be presented here as well as others you may come across. In addition, it will be of help should you want to design your own program later on.

selecting an exercise program

purpose of the program

A question is often asked, "What kind of exercise is best?" The answer, of course, depends on what your goals are. What do you want to accomplish? Only when we have this answer can the program and specific methods be determined. The exercise program must be specific to your particular needs or desires.

In general, the purposes may be: (1) corrective; (2) fitness for normal living; or (3) fitness for a specific task or sport. Should the purpose be corrective, such as regaining strength in an injured knee or correcting certain postural defects, special exercises and methods would be called for. Conditioning for a specific type of athletic competition would require an analysis of the major components necessary for top performance and intensive training to develop them rapidly to the highest degree possible. This often involves a program employing specialized equipment and methods not appropriate for those concerned only with reaching and maintaining a minimal level of fitness for health, productive work, enjoyment of recreational activities, and emergencies. Fitness for normal living requires equal attention to all the components measured in the fitness evaluation, conducted with less urgency and at a less intensive level.

Though the principles to be discussed apply to all exercise programs, our focus will be on programs for normal living. In this respect, certain factors make some types of activities and exercises more suitable than others.

limitations and uses of competitive sports

For various reasons, it is not advisable to rely solely on competitive sports to achieve and maintain a desirable level of physical fitness. The seasonal nature of sports, the requirement of special equipment and facilities, and the need for teammates and/or opponents are all factors that can limit opportunity for the regular year-round participation needed for fitness. The time consumption factor in sports activity is another limitation. Training for improvement requires at least three days a week participation and preferably more. A minimum of three days a week is required just to maintain a desirable level of condition. When this number of days is multiplied by the hours usually required for most sports activities, the total time required is often more than the average person can afford. In addition, regular conditioning and competition is essential to avoid recurring muscle soreness or possible injury. Though it may be possible to overcome these limitations for a time, eventually increasing age will force discontinuance in competitive sports. The need for physical fitness, however, will still remain.

Finally, one of the major drawbacks is the inability of competitive sports to develop all motor fitness elements and cardiovascular endurance equally or to a sufficient degree for improvement or maintenance in all areas. Each sport tends to stress only certain elements. Even athletic teams find they must engage in a special and continuous conditioning program not only to prepare for their activity, but also to maintain their high level of fitness during the season.

For some, provided there is opportunity, competitive sports may be used to supplement a maintenance program once a desired level of fitness is reached. Certainly many can be used for recreational purposes. For certainty in training, however, you should have a knowledge of one or more balanced exercise programs. Knowing more than one can help to avoid boredom and monotony over long-term use and insure the regular participation so essential for physical well-being.

criteria for exercise training programs

The program selected should meet all or most of the following criteria:

1. Require no specialized equipment or facilities.
2. Be able to be performed regardless of where you are.
3. Require little time.
4. Require no other participants.
5. Be adaptable to any age or physical condition.
6. Develop all elements of cardiovascular and motor fitness and be adaptable to other fitness components.
7. Exercise all major muscle groups of the body to insure symmetrical (even) development.

special conditioning activities

In addition to programs which meet the above criteria, a knowledge of certain conditioning activities which require some specialized equipment and, perhaps, a little more time is also desirable. Such activities as weight training or the Exer-Genie Exerciser program can add variety to your conditioning program, help speed up development in certain areas, or meet special needs.

basic principles

adaptation

The adaptation principle means that the body will respond and adapt specifically to the kind and amount of the physical demands placed upon it. When faced with the stress of exercise, the body will immediately mobilize its resources to meet the demands: increased heart rate and respiration, blood flow to the working tissues, etc. As we have seen, it is this quality and the repeated forcing of the body to adapt to higher levels of stress that brings about the beneficial long-term effects of exercise. But the body will also adapt to particular patterns of use or disuse.

For example, if the body is asked to perform only normal daily activities over a long period of time, it will adapt only to the level of these demands and be unprepared for anything more strenuous. In the same manner, it will respond to an arm immobilized in one position for a long period of time (such as when in a cast) by a decreased blood supply to the arm, resulting in a loss of muscle strength and tone. The arm has to be worked again to regain its strength, tone, flexibility, and full range of motion. Another example of specific adaptation to a pattern of use is when an arm is used constantly through only a partial range of motion. Only those tissues involved in the movement receive the increased blood supply. Over a period of time, the muscles and tendons adapt to this restricted movement by shortening, thus making further range of motion difficult.

Excluding permanent physical defects, your physical condition is not static, but reflects the specific adaptation of your body to the everyday demands placed upon it. If the demands are low, the level of adaptation will be low. If no greater demands are made, adaptation that results in improved physical fitness cannot occur.

overload

The overload principle is the key to improving your physical condition through exercise. Based on the adaptation principle, the overload principle is simply this: In order to improve any physical function, it is necessary to place that function under a repeatedly greater than normal work load until it has adapted to the increased demand. A corollary to the principle is that the degree of improvement is directly proportional to the degree of the overload.

applying the adaptation and overload principles

In applying the principles of overload and adaptation to a training program, there should be three stages, each guided by other specific principles. The first is the toughening stage and the guiding principle here is a moderate beginning. The second is the slow improvement, or developmental stage, guided by the principle of gradual progression. The final stage is the maintenance or sustaining stage when overload is discontinued and regularity and diversity become the guiding principles.

the toughening stage (moderate beginning)

Though the body responds immediately to meet the demands of exercise, it requires some time to become accustomed to handling higher exercise loads. The beginning of a training program will usually be characterized by stiffness and soreness. The degree of distress will depend on your present level of condition and the intensity of the increased demands. The soreness is caused by an inadequate blood supply to the untrained working muscles and the build up of lactic acid in the muscle tissues. The lactic acid is very slowly removed during the recovery period and its prolonged presence irritates the nerves of the muscle fibers.

To avoid undue soreness and fatigue at the start, care should be taken that the overload is applied moderately and geared to your current physical condition. For those unaccustomed to exercise, those who have been inactive, older persons, or those recovering from illness, a rapid overload can be dangerous. At best, it can cause such discomfort that a negative attitude towards physical conditioning may result. After completing a workout and tapering off, you should have a pleasantly tired and relaxed feeling. Excessive fatigue and soreness which disturbs sleep or carries over to the next day is an indication of "straining" rather than training. On the other hand, its complete absence would indicate an inadequate overload.

Some soreness and fatigue is necessary in order to "toughen" the body and prepare it for gradually increasing overloads and improvement. This toughening stage will usually last about two weeks, varying more or less depending on individual condition and the intensity of the exercise. During this period, because of general fatigue and soreness, there may actually be a retrogression in most measures of physical performance. When this happens, some people become discouraged, and a natural tendency to give up the effort must be fought. As exercise is continued, blood supply through the tissues increases and becomes more efficient, both during the exercise and recovery period. Waste products build up more slowly and are removed more easily. After passing through the toughening stage, adaptation to increased demands is made with very little or no distress, provided the overloads are applied gradually.

the slow improvement stage (gradual progression)

Once the initial period of adjustment is over, improvement of the various systems can be made only by progressively increasing the exercise load over a period of time, until the desired level of improvement is reached. A gradual progression is necessary to avoid a recurrence of soreness and fatigue. One or a combination of several methods may be used to increase the workload depending upon the nature of the program.

Increase the resistance. (1) Increase the weight that must be handled in a particular exercise, such as in weight training; (2) make a specific rapid body movement more difficult to execute, such as jumping or running with weighted boots, running uphill, or swinging a weighted tennis racket; (3) increase the rate of work, thus enabling inertia to increase the resistance. For example, perform a given amount of work in a shorter period of time, such as running a mile or performing a specific number of push-ups in less time; or perform more work in a given period of time, such as increasing the number of sit-ups done in one minute from twenty to thirty; or exert a greater force against an immovable object for a certain period of time. Overloading rapid body movements or increasing the rate of work are referred to as *speed overload techniques.*

Increase the duration of the exercise. (1) Increase the number of repetitions of a given exercise while maintaining the same rate, such as in weight training or calisthenics; (2) increase the distance covered while maintaining the same rate, such as in running a mile-and-a-half instead of a mile; (3) cut down the number of rest periods during a workout; or (4) lower the actual resting time during each rest period.

Improvement during the developmental stage is fairly rapid at first, but becomes progressively slower as the body nears the upper levels of its capacity. At times, certain plateaus will be reached which call for a temporary break in the normal pattern of progression and a considerable increase in the exercise load in order to make further improvement. These plateaus become more frequent and difficult to break as the body nears its maximum level of performance. Improvement becomes less noticeable. When this point is reached (or when preestablished goals are reached), the developmental program should be ended and the maintenance program begun. The slow improvement or developmental period can last anywhere from four to sixteen weeks, depending on age, starting level of condition, rate of progression, and type of program.

the maintenance stage (regularity, diversity, discontinuance of overload)

After reaching the point where little improvement can be made (or no further improvement is desired), the objective then is to maintain the level of fitness reached. Overload is no longer applied and the frequency of exercise may be reduced to a minimum of three days a week. Provided adequate supplementary exercises are performed, the emphasis during this stage may switch more to games and sports for the additional enjoyment and benefits that may be obtained. The maintenance program should be made very interesting, yet continue to call for an adequate amount of vigorous activity.

related principles

rate of improvement and intensity of the exercise

Though improvement can occur only when the work load is increased, the rate of improvement is more directly proportional to the intensity of the overload rather than the amount of the work increase itself. That is, the greater the intensity, the faster the gain. By intensity we mean the degree of force or how strenuous the overload is.

For example, using the physics formula: Work = Force × Distance, we can see that the total amount of work done can be increased by either increasing force (amount of resistance: weight, rate, etc.) or distance (actual distance, repetitions, etc.), or both. Lifting a 10 pound weight 50 times over an equal distance of 2 feet, for example, would be twice the workload (1,000 ft./lbs.) of lifting 50 pounds 5 times (500 ft./lbs.). But the intensity of the work, in terms of force and effort involved is obviously greater with the heavier weight. Though both procedures might constitute an overload for a particular individual and result in improvement, the improvement would be faster using the heavier weight, even though the total workload was lower. Similarly, running a mile in 8 minutes is a more intense overload than running 2 miles in 20 minutes and even though the total workload might be lower, a faster rate of circulorespiratory improvement would occur.

Precautions. The same precautions are necessary in determining the intensity of the exercise program as in applying the general overload: moderate beginning and gradual progression. For many older persons or those with certain medical problems, it is better to apply the overload at a low intensity level. Though improvement will be slower, it will still occur, and the safety advantages far outweigh the slower progress.

retrogression

This principle was touched upon previously, but is repeated in order to specifically emphasize that before improvement begins, whether in a conditioning or skill development program, you can expect your performance to be temporarily worse than when you began. This is a necessary part of the adaptation process and nothing to become discouraged about. By persevering, the body will adapt to the new demands and improvement will begin.

specific effects of exercise

The effects of exercise are specific to the type of activity engaged in and the particular body function it exercises. That is, if your objective is to improve strength, then exercises which call for progressively increasing applications of strength must be used. In general, to improve muscular endurance, the emphasis must be on repetition overloads, using lighter resistances which will allow for at least a certain number. Similarly, flexibility will only be improved by exercises which demand increased ranges of movement and improvement of circulorespiratory endurance would call for activities which place prolonged demands on the heart and lungs.

Identification of muscle action. Because of the specific effects principle, the ability to identify muscle action can be helpful in determining the particular muscles affected by various exercises or in selecting an exercise that will work on a specific muscle or muscle group that needs improvement. A simple technique can be used for this purpose.

To determine the specific muscles affected by a particular exercise, perform the movement isometrically (against some immovable object or a partner offering resistance) and then note the muscles that come under tension. The muscles that feel the hardest are those which are being worked most. Those that are relatively soft and loose are not being worked. To select the best exercise for a particular muscle, perform a similar isometric contraction in every possible direction at the nearest joint or joints to the muscle and feel the muscle. The movement which results in the firmest contraction is the one that should then be performed and overloaded accordingly.

Example of specific effects principle. The degree to which the specific effects principle operates was demonstrated in a study involving a group of university students.[1] On a pre-test, an experimental group was run to exhaustion on a treadmill set at both a moderate intensity (7 m.p.h.) and high intensity (10 m.p.h.). After running three miles a day for several months, the group was retested. Although they ran more than four times longer at the moderate intensity level, their ability to run at

---

[1] Wayne Van Huss, et al., *Physical Activity in Modern Living* (2nd ed.) (Englewood Cliffs, N.J.: Prentice-Hall, Inc., 1969), p. 47.

the higher intensity level actually decreased. Their improvement was in the area most closely related to the type of training.

The specific effects principle explains why even "highly conditioned" athletes find that they are not "in shape" when they switch from one sport to another, such as from football to wrestling or from cross-country to tennis. It also explains the slight muscle soreness experienced by persons who switch from a jogging program for general endurance to a running in place program. The significance of the principle is that "you get only what you train for." For desired results, you must train specifically! Because of this, a well-rounded physical fitness program must include a variety of exercises and activities selected carefully according to your objectives.

adjusting and evaluating the program individually

What might be considered another basic principle is that an exercise program must be fitted and evaluated on an individual basis. Individual differences must be considered in determining the starting point, rate of progression, and the level to be reached. The program should never be viewed as competitive. Though there are wide variations within similar groups, the following factors should be considered.

age

The period of maximum physical maturity is between the ages of twenty-five and thirty. After this point, there is a slow decline in maximum strength, increasing after the age of fifty, but even at sixty the loss does not usually exceed 20 percent of maximum.[2] Decline in speed of movement and reaction time shows a similar pattern. The loss in other areas of motor fitness is probably similar, but coordinated functions involving several elements appear to decline more quickly with age.

Maximum heart rate, cardiac output, and maximum oxygen consumption also show a general decline with age, with cardiac output at rest ($SV \times HR$) generally showing a decline of about 1 percent a year after maturity. In addition, there is a slower heart rate and blood pressure response to stress.

These factors all combine to make improvement slower and to a lower level, and recovery from stress slower than at an earlier age. In addition, deconditioning is faster and to a lower level, and the reconditioning

[2] Herbert A. de Vries, *Physiology of Exercise* (Dubuque, Iowa: William C. Brown Company, 1966), p. 284.

process slower. Overload and intensity, therefore, must be adapted accordingly. For older persons, also, strength, speed, and agility take on relatively less importance than flexibility, local muscle endurance, body weight, relaxation, and working capacity—and this should be reflected in the program. There is no evidence, however, that vigorous exercise can injure a healthy individual, regardless of age. In fact, if regular exercise is continued, an older person is often capable of more vigorous activity and can maintain a higher level of condition than a much younger person.

sex

Strength differences between men and women appear to be caused basically by differences in muscle size and cultural patterns (tendency towards overprotectiveness and less physical activity after puberty) rather than by differences in muscle quality or sex per se. Generally speaking, because of lower muscle quantity and smaller muscle size, the strength of women is lower than in men, though a woman with larger muscles than a man can certainly be stronger. Most evidence shows that there is little difference in muscular endurance if the strength factor is ruled out. Women, in general, are more flexible than men at all ages. Because of a naturally greater amount of fat tissue than men, women need not fear the development of bulging muscles as a result of exercise. In fact, it should be remembered that beneath every curve, there is a muscle!

In terms of circulorespiratory function, we know that the heart rate for women is from five to ten beats per minute faster than for men at all ages. Before puberty, however, no significant difference has been found in maximum oxygen consumption. After puberty, working capacity has generally been found to be about 85 percent that of men in similar age groups. How much of this is because of smaller heart muscle size and strength capacity and how much from decreased activity caused by cultural patterns after puberty would be interesting to determine. Many physiologists believe that cultural differences play the more important role.

health status

Obviously a person's health status is of primary concern in determining the amount and intensity of the overload and the rate of progression. For those attempting to improve heart and other cardiovascular problems, close supervision of the program by a physician is essential. Frequent medical examinations (at least yearly) are desirable, especially in middle-age, for protection against overstress during the early stages of an unsuspected illness.

Overload and intensity should be reduced during even a minor illness, such as a cold or minor infection or injury, since the body under such conditions is already under greater than normal stress. If necessary, the program should be discontinued for a time. In resuming the program following an illness or temporary halt for whatever reason, it is safer to start again at a lower workload and begin the gradual progression again.

present level of condition

Your level of condition at the start of the exercise program will not only determine the beginning level of exercise, but be reflected in your rate of improvement. The less fit you are, the greater the improvement that will be shown on the basis of actual measurement scores. Conversely, the closer you are to your maximum performance level, the less will be shown and the more effort that will be required to make gains. Relatively speaking, for example, reducing the running time for a mile from 6 minutes to 5:45 would be a more noteworthy accomplishment than reducing it from 10 minutes to 8 minutes.

previous activity and conditioning experience

Though there are no specific residual benefits to exercise once a program is discontinued, the person who has had previous training experience and at one time in his life has reached a high level of fitness will respond more quickly to training than one who has not. Not only will he respond more quickly, but he will usually achieve a higher level of fitness or skill. Mesomorphs, medials, and combinations of these body types will also usually respond more quickly to a training program than either pure ectomorphs, endomorphs, or combinations of them.

psychological factors and motivation

Each person enters a training program with certain built-in feelings and perceptions regarding exercise, his level of skill, ability to learn, and ability to endure a particular training procedure. In many cases, the amount of improvement is limited more by these psychological factors than by actual physical capacity. More often than not, the urge to discontinue exercising will result not from physical exhaustion, but from a self-imposed psychological limit. This must be recognized and fought. Motivation, based on preestablished minor and major physical or psychological objectives (exercise repetitions, weight loss, self-esteem, etc.), is an essential ingredient for success in the program. By the same token, a person's psychological "drive" may exceed his physical capacities. This also must be recognized and curbed to avoid harmful overexertion.

goals and measurements

Predetermined goals provide not only the motivation, but also the basis upon which the type and amount of exercise is determined and progress is measured. Without them, a person doesn't know where he's going, how to get there, or when he has arrived. The goals should involve both specific fitness component objectives (specific heart rate, body weight, etc.) and specific exercise objectives, such as a certain number of repetitions of a particular exercise or running time for a certain distance. Intermediate objectives and periodic testing along the way help to determine progress, make necessary adjustments in the exercise prescription, and reinforce motivation.

To determine the appropriate type and amount of exercise, we must know: What needs to be improved? What needs to be maintained? What levels are to be reached? The fitness evaluation gave us the answers. The types of exercise used must be those which work specifically on the areas that need development or maintenance (specific effects principle). The amount needed must be greater in those areas that need improvement rather than those in which maintenance is the goal. Now we will examine the types of exercise there are, what they are best for, and the specific methods for improving the various components of fitness.

general types of exercise

Though difficult to separate completely, exercises may be classified generally into four major types. Each has specific effects, involves different procedures, and requires a varying amount of equipment and time. Choice depends on these factors in relation to purpose, objectives, and personal enjoyment preference.

isotonic exercises

Isotonic exercises are those involving muscle contraction that produces movement through a partial or complete range of motion. Technically speaking, all activities that involve movement are isotonic, but the term is used primarily in referring to activities such as weight training and calisthenics. These exercises are associated mostly with the development of muscle tone and other elements of motor fitness: muscle strength, endurance, power, flexibility, balance, agility, coordination.

isometric exercises

Isometric exercises involve the static contraction of a muscle or muscle group; that is, contraction in a fixed position which produces no movement. Examples are tightening the fist, flexing the biceps muscle, pressing the palms together, or pushing against an immovable object, such as a wall. The primary use of isometric exercises is the development of muscle tone, size, strength, and power. Since blood flow through the muscle tissues is restricted during an isometric contraction, there is little effect on muscle endurance and none on general endurance.

anaerobic activities

Anaerobic activities are those which require so much oxygen that a rapid oxygen debt is created and the activity cannot be continued for any extended period of time. Though isometric contractions are anaerobic in nature, primary examples are short distance, high speed running or swimming, weight training, and heavy calisthenics or gymnastics. Such activities are used primarily to build up speed, power, and muscular strength, including heart stroke volume, and in training to build up anaerobic capacity needed for vigorous sports competition. Activities which require reasonable amounts of oxygen, such as light calisthenics, but which are not carried on long enough to produce a good overall training effect on the pulmonary and cardiovascular systems are also classified as anaerobic.

aerobic exercises

Aerobic exercises are those which demand oxygen without creating a large or rapid oxygen debt and which can be continued for a long enough period of time to produce a good overall training effect on the lungs, heart, and cardiovascular system. Long-distance running, swimming, walking, and cycling are included in this category, along with running in place, rope jumping, handball, tennis, and similar activities. In addition to their effect on the circulatory and respiratory systems, aerobic activities have a good effect on muscle tone, muscle endurance, body weight and proportions, agility, and coordination.

strength

Strength improvement is best brought about by progressively increasing the resistance against which a particular muscle or muscle group must work. Either isotonic, isometric, or a combination of these exercises can be effective for improving both strength and muscle tone.

Weight training. Weight training is considered the fastest and best method for improving strength because of the ease in measuring and controlling the exercise load and because the muscles are worked through a complete range of motion. In weight training for strength, particular lifting movements are performed using relatively heavy weights and a few number of repetitions maximum (RM). *Repetitions maximum* means the maximum load that can be lifted through a full range of motion a designated number of times. Various studies have shown the greatest gains coming from progressive load increases in the low RM ranges. An optimum appears to be three sets of six RMs. A *set* refers to the number of times a given number of exercise repetitions are performed. Detailed weight training principles and procedures for developing strength, as well as power, muscle endurance, and hypertrophy will be covered in the programs chapter which follows. Hypertrophy refers to an increase in muscle size—but not necessarily with a corresponding increase in strength.

Calisthenics. For general purposes and ease of performance, calisthenics remains a popular method. In calisthenics, resistance is increased by speed overload: performing more repetitions in a given time or the same number in a shorter time. Since maximum loads cannot be achieved, however, maximum strength gain cannot be developed. Furthermore, since the method of increasing resistance involves increasing rate and repetitions, the elements of speed and endurance become larger limiting factors.

Isometric exercises. Significant improvements in muscle size, static strength, and dynamic strength have been reported from the use of isometric exercises. Effective development, however, requires daily maximum contractions held for at least six seconds. Higher strength gains have been reported by increasing the number of repetitions to between five and ten times a day, with four or five workouts a week appearing to be most favorable. It has also been reported that the strength of a particular muscle group can be retained by as little as one maximum contraction a week. Since strength gains are specific to the angle at which the resistance is met, however, it is usually recommended that the static contractions be applied at a minimum of three points through the entire range of motion of a particular muscle for best results. A basic program for general development and maintenance is included in the programs chapter.

Conflicting reports and controversy still exist over the relative merits and disadvantages of isometric exercises. These will be discussed further in the programs chapter and later in this chapter, but one criticism is that since range of motion is restricted, strength gain is not as effective as with isotonic exercises. Another is that the lack of movement impairs flexibility and relatively weakens the tendons and ligaments at the joints. In order to retain the major advantage of isometrics (maximum strength gain in a minimal amount of time), but eliminate these disadvantages, efforts were made to devise an exercise procedure that would combine the advantages of both methods.

Combination isometric and isotonic exercises (iso-kinetics). Probably the most popular device combining isometric and isotonic techniques is the Exer-Genie Exerciser. The Exer-Genie Exerciser is a mechanical device that allows for performing a series of exercises in which first a static contraction is held for ten seconds and then is followed by a full isotonic contraction against a steady resistance. The exercises are done by pulling, pushing, or lifting against a bar or strap attached to a nylon cord encased in a small cylinder device which is stabilized and can be set to control the amount of resistance against the movement of the cord. Details of the program are given in Chapter 5.

power and muscle endurance

Training for power and endurance is closely related to the types of exercises and procedures for developing strength, but with some variations. Normally, they develop simultaneously with strength, but specific techniques can be used to develop them more effectively.

Power. Power refers to the explosiveness of body movement or the rate at which force is produced. As such, it involves two elements: strength to produce the force and speed to increase the rate at which it is applied. Thus, power can be increased by either improving strength or speed of movement or both.

Since they involve these elements, speed overload techniques are the most direct methods for developing power. This may be done by overloading a specific rapid body movement, such as jumping with weighted boots (the most direct method) or by increasing the rate of work. Again, because of the greater loads possible, increasing the rate of work in a weight training program would produce greater results than increasing the rate of calisthenics.

Training for improvement in speed of movement involves simply the effort to gradually make the movement more quickly and to eliminate unnecessary movements. Such training, a separate one in itself, results in a rapid early improvement, but to a plateau from which it is difficult to improve, as experience can testify. At this point, further improvement in power depends primarily on improvements in strength. Since very few persons even begin to approach their maximum strength, large gains in power can be made this way.

For activities in which specific power movements at certain points are important, such as getting off the mark in track or at the point of contact in hitting a baseball, isometric contractions at these points can be an effective method. The Exer-Genie Exerciser method is also effective for this kind of power development.

Muscle endurance. For local muscle endurance, the duration of the exercise must be increased, but for the greatest gains the resistance must be high enough to constitute an intensity overload as well. Merely increasing the number of workouts or extending the workout time while using very light resistance exercises will result in very little increase. In weight training, concentration on the development of muscle endurance would normally involve increasing the number of RMs to no more than twenty-five. Muscle endurance, as well as strength, can be increased also by adding to the number of sets performed or decreasing the resting time between exercises and sets. The same principles would apply to a calisthenics program.

Since by its nature isometric contraction cuts off blood flow and depends for its continuance only on the energy supply available, there is no way that these exercises can increase muscle endurance. Muscles trained by isometric methods only usually fatigue easily when subjected to prolonged work.

flexibility

Flexibility refers to the range of motion possible at a joint or series of joints, such as the spine. As such, flexibility is specific to a particular joint or combination of joints and it would be inaccurate to speak of persons as being either "flexible" or "inflexible." At some joints the degree of movement is very definitely limited by the bone structure itself or the bulk of the muscles between the joint. Flexion and extension at the elbow and knee joints are examples. Extension (straightening or increasing the angle at a joint) is definitely limited by the joint structure; flexion (bending or decreasing the angle at a joint) by the size of the biceps, forearm, and calf muscles. At such joints as the ankle, hip, and shoulder, however, flexibility limits are set primarily by the ease and degree to which the muscles, connective tissue (tendons and ligaments), and skin can be stretched.

Types of flexibility and significance. There are two types of flexibility: *static* or extent flexibility and *dynamic* flexibility. Static flexibility is the ability to stretch a body part as far as possible in various directions, but is not necessarily a good indication of the "looseness" or "stiffness" at the joint. Dynamic flexibility involves the ability to make repeated rapid movements at a joint with little resistance to the movement. The development and maintenance of both static and dynamic flexibility is important for graceful movement in walking and running, preventing joint pains and backaches, and for most sports activities. In some activities, such as swimming, diving, and tumbling, flexibility is all-important. Poor flexibility results in bound muscles and stiff, awkward movement. For physical performance, dynamic flexibility is probably more important than an extreme degree of static flexibility, since there is no particular advantage in the latter.

Training for flexibility. There are two kinds of stretching exercises by which flexibility can be improved or maintained: *static* and *ballistic*. Static stretching involves locking the joints in a holding position, stretching the muscles and tendons to the greatest extent possible, and holding the position for a period of time, usually five to ten seconds. Ballistic exercises (typical calisthenics) involve slow bobbing and bouncing movement whereby muscles are stretched by momentum at the end of the range of motion.

Both methods are equally effective in improving static flexibility, but it is logical that ballistic stretching would have a greater effect on dynamic flexibility. In performing both, however, care must be taken that they are done at a slow to moderate pace and without any "jerky" movements. The reason is that when a muscle is stretched, a reflex contraction results in it roughly proportional to the amount and rate of the stretching. This is overcome in part by another reflex action which inhibits the force of contraction in the entire muscle group being used as a protection against overstress. Nevertheless, in quick, jerky movements, the contracting force may be large enough to counteract most of the benefits gained by the stretching. Such movements can also cause muscle soreness, a knotted or pulled muscle, and there is the added danger of possibly tearing muscle and tendon tissues. In this respect, static stretching offers an advantage in that there is less danger of muscle soreness or tissue tear. In addition, in static stretching, the inhibitory reflex operates at a higher threshold of stretch than the contracting reflex, thus allowing a greater stretch which can actually relieve soreness. A sound procedure in warming up cold muscles or improving flexibility is to precede ballistic stretching with brief static stretching at the various joints.

Though calisthenic-type ballistic stretching exercises or static stretching exercises are best for developing flexibility, weight training can also be effective provided exercises are performed through a complete range of motion at all the joints involved. If this is done, there is no evidence that weight training can harmfully affect flexibility. Supplementary exercises should be included, however, for joints and movements not affected by or included in the weight training program.

speed, agility, and coordination

Basically, speed depends on the relationship between the driving force that provides the movement and the forces that resist the movement. To improve speed, then, the positive forces must be increased and the negative forces decreased. The most important positive factor is strength. Speed can be improved by strengthening the muscles involved in the particular movement for which speed is desired, preferably by dynamic movements closely related to the skill. Negative forces can be reduced by improving flexibility and coordination, analyzing the skill and applying forces correctly, and eliminating all unnecessary movements. In running, air resistance may be reduced by running in a crouched position, but the mechanical disadvantages may far outweigh the benefits gained. Of course, maximum intensity is essential, but it should be preceded by a thorough warm-up that produces sweating. Such a warm-up increases the speed of muscle contraction and relaxation and can help to avoid soreness and strain.

Speed capacity appears to be limited by intrinsic differences in the speed of muscle contraction (varies inversely with the size of the muscles in animal species) and by the quality of neuromuscular patterns (coordination). Like other elements, speed of movement is a specific rather than a general quality.

Agility, or the ability to change direction of movement quickly, involves the elements of speed, strength, and coordination. Training should include development of speed and strength, as previously described, and activities that develop coordination, such as calisthenics which involve simultaneous multiple limb movements. Specific agility training would involve making a series of rapid total body movements involving quick changes of position and direction in rapid succession. Again, these qualities are specific to the type of activity engaged in.

general (circulorespiratory) endurance

To improve general endurance, aerobic activities are essential. Anaerobic activities have little effect on improving general endurance, because the oxygen demands are either not great enough or so great that a rapid oxygen debt is created and they cannot be continued long enough to produce a training effect. For improvement, the oxygen demands must be great enough and long enough to produce a good training effect on the heart, lungs, and blood vessels. The aerobic activities provide this by demanding a continuous supply of oxygen without building such a rapid oxygen debt that the activity must be cut short.

Though aerobic sports activities (handball, tennis, etc.) can be used, continuous-type activities such as walking, running, running-in-place, and swimming are safer and easier in terms of controlling the workout and measuring progress. Switching to a sport activity can be done later, for maintenance, after a good level of condition is reached. A jogging or running program beginning with alternate run-walk intervals is probably the most adaptable and convenient method of initial training. For those who have been inactive for some time, it should be started by walking in order to condition the feet and legs.

Producing a training effect. The training effect essentially depends on how high the heart rate is elevated and how long it is kept there. The best study on the minimum heart rate and the length of time it must be sustained to produce a training effect is, in our opinion, one made by Dr. M. J. Karvonen,[3] because it can be used to truly individualize a training program. In studies on untrained medical students, Karvonen found that to improve cardiovascular function, the heart rate during exercise must be increased to at least 60 percent of its "Working Heart Rate" and be maintained there from several minutes to fifteen minutes or longer. The higher it is elevated and the longer it is sustained, of course, the greater the effect. The *Working Heart Rate* is the difference between an individual's Resting Heart Rate and his Maximum Heart Rate.

Determining the minimum exercise heart rate. It has been well established that the Maximum Heart Rate decreases with age and that heart rates for women are from five to ten beats higher than for men at all ages. Though there are variations in each age group that make it possible for a person of sixty to have a higher Maximum Heart Rate

3 M. J. Karvonen, "Effects of Vigorous Exercise on the Heart," in *Work and the Heart*, eds. F. F. Rosenbaum and E. L. Belknap (New York: Paul B. Hoeber, Inc., 1959).

than one of thirty, a conservative formula for estimating Maximum Heart Rate is 220 minus age for men, and 227 minus age for women.[4] The most accurate Resting Heart Rate would be obtained by taking a one-minute pulse count before arising in the morning and averaging it out over several days. To determine the minimum Exercise Heart Rate needed for a training effect the Karvonen formula is: EHR = Max HR − Resting HR × .60 + Resting HR. For example, taking a young man, age twenty, with a Resting Heart Rate of 70, his minimum Exercise Heart Rate would be $200 - 70 = 130 \times .60 = 78 + 70 = 148$. To improve his general endurance, he would have to bring his heart rate up to at least 148 and sustain it for at least three minutes or longer.

From other studies and what we know about overload and adaptation, however, it is logical that any aerobics program that lasts from ten to twenty minutes and sustains a heart rate above normal (even if slightly below the Karvonen 60 percent figure) would produce some training effect, no matter how minimal. The lower the Exercise Heart Rate, of course, the slower the progress (Rate of Improvement–Intensity principle) and the longer the duration of the exercise must be. By the same token, sustaining a higher EHR would produce a faster effect and require less workout time. Cooper gives as examples of equivalent training effects either maintaining a 150 heart rate for ten minutes or a 130 for forty-five minutes.[5] For those in poor condition and all older persons, the lower intensity is a safer procedure.

Setting the limits. In any event, it must be emphasized that it is neither necessary nor desirable to sustain a Maximum (or even near maximum) Heart Rate to produce a beneficial effect. Though many athletes do reach these rates for periods of time during competition, it must be remembered that these persons are usually in top condition. For the average young person and all older persons interested in general conditioning it is an unnecessary practice. For the middle-aged or older it can even be dangerous.

For both effectiveness and safety, what we like to do is establish an Exercise Heart Rate Range that a person attempts to keep within during his training program, and to use a system of self-heart-monitoring to see that he stays within it. The monitoring is done by taking quick ten-second pulse count checks at intervals during the training session or immediately following it, similar to what we did in the step test during

---

[4] The formula 220 minus age was given at the San Diego State College Symposium on Adult Fitness, June 1970. It was adapted to 227 minus age for women based on the established higher heart rate for women.

[5] Kenneth H. Cooper, M.D., *The New Aerobics* (New York: M. Evans & Company, 1970), p. 155.

the fitness evaluation. It gives a truer measure of the intensity of the workout for a particular individual than using either total running distance or time alone, and allows him to continually adjust the workout to assure that he neither under- or overexercises. Thus, the program is truly individualized to provide an effective and safe rate of progression.

Establishing the exercise heart rate range. Using the Karvonen formula for those in poor condition and all older persons, we will set a minimum and maximum range based on the 60 and 70 percent figures respectively. The exercise bout should then be set and adjusted so that the heart rate reaches and is sustained for at least a few minutes and up to fifteen minutes or longer at the minimum figure determined, but not higher than the maximum rate calculated. At the start, the exercise bout may be a walk, an alternate jog-walk, or a steady jog for a distance of one mile. Based on actual heart rates reached, the pace and method of covering the distance are then adjusted in subsequent workouts. As training progresses and the heart and related systems become stronger, a greater and greater exercise load (intensity and/or duration) would be required to maintain the same exercise heart rate range—a sign of improvement. After several months, the range may be increased to between 70 and 80 percent of the working heart rate for faster and greater improvement.

For young college-age adults, unless in extremely poor condition, a beginning range based on 70 and 80 percent calculations may be used for faster progress with little or no danger.

Monitoring the heart rate. By dividing the one minute minimum and maximum Exercise Heart Rates by 6, a range based on ten second pulse counts can be determined. For example, a minute range of 152 to 162 would mean ten second counts of between 25 and 27. Because the ranges are set conservatively, early errors in counting and "accidental overexertion" should not be alarming. Using an alternate jog-walk procedure as an example, let's see how the monitoring would work.

You would start off at a slow to moderate jog and when breathing starts to become difficult, stop, locate the pulse, and while continuing to walk slowly, take a ten-second count. While walking, it is easier to take the count at the neck, temple, or over the heart. If it is above 27, make a mental note of your rate and depth of breathing so as not to allow it to get that high again during the next running interval. Continue to walk slowly and take ten second counts. Note your breathing and how you feel when the count reaches 27 and use it as a future guide. When the count reaches 25 or sooner, resume jogging until the next check interval. When the point of a steady jog or run is reached, or for those who can start with it, the count should be taken immediately after the run with a note made of the time for the distance covered, so that the pace can be adjusted in future runs. It has been our experience working with college-age, middle-age, and older adults that after a week or two, the lower and upper limits can be judged quite accurately even without a pulse count.

The maintenance goal. The question now is at what point does the training for improvement cease and the maintenance stage begin? Basically, the answer is: when your Resting Heart Rate goal and maximum oxygen consumption goal (as per the 12-minute test) is reached. Cooper sets the maintenance goal for the average person in terms of distance and time for a particular kind of aerobics program performed a certain number of days a week. For a running program, among the minimum goals is any of the following:[6]

  1 mile in under 6:30 minutes, 5 times a week.
  1 mile in between 6:30 and 7:59 minutes, 6 times a week, or
  1½ miles in 12:00–14:59 minutes, 5 times a week.
  2 miles in 20:00–23:59 minutes, 5 times a week.

Alternate goals for walking, stationary running, cycling, and other aerobic activities are also given in Cooper's books. Application of the Exercise Heart Rate monitoring technique can be effectively used in all these programs to individualize them.

training for other fitness components

posture

Both isotonic and isometric exercises are used to correct postural deviations. The basic principle underlying all exercises designed to correct deviations caused by muscular imbalance is: "Stretch the Short Side and Strengthen the Long Side." For example, if the deviation is a shoulder tilt to the right, the side muscles on the right side will be shorter than that the muscles on the left. To regain the muscular balance of forces, the right side must be stretched by flexibility exercises and the overstretched muscles on the left side shortened by strengthening exercises—working them in side flexing movements against increased resistances. Once balance is regained, excluding any structural deviations,

---

[6] From *Aerobics* by Kenneth H. Cooper, M.D.; © 1968 by Kenneth Cooper and Kevin Brown; page 40. Adapted by permission of the publisher, M. Evans and Company, New York, New York.

good posture can be maintained by a balanced exercise program that works and stretches all major muscle groups through their full range of motion. Being posture conscious will also help to correct or prevent the gradual muscular adaptation to repeated poor posture positions. Standards for good posture and specific corrective exercises are covered in Chapter 6.

body weight and body proportions

An exercise program aimed at losing weight and unsightly fat must include activities which expend the most calories. Essentially, these include the aerobic activities, but certain vigorous anaerobic activities such as wrestling also call for high energy expenditures. Fig. 11 in Chapter 7 lists the calorie cost of various activities and Chapter 7 explains how to construct an exercise-diet program for the best results.

Stretching and light exercises with a high number of repetitions can also help body proportions by stretching the skin and muscle tissue and increasing muscle tone. Vigorous one to two minute finger and thumb massage of thick, lumpy fat areas (fibrosis) can help to loosen the skin and increase circulation in the surrounding tissues, thereby making the energy breakdown of fat reserves more efficient.

relaxation

The degree of muscular tension present in an individual is closely related to the degree of mental, emotional, and psychological stress he experiences. When we say someone is "tied up in knots" or "tight as a drum," we're referring to nervous tension, but it can be actually seen in the tenseness of his muscles. Conversely, the conscious release of muscular tension through certain techniques or from a vigorous workout has a reciprocal effect on nervous tension from psychological or emotional stress. Application of this principle and exercise techniques for the conscious release of muscular tension is discussed more fully in Chapter 8.

other exercise guides and precautions

time of day to exercise

Vigorous exercise interferes with the digestive process for approximately one hour before or after the intake of food. It is best to delay exercise at least 1 to 1 1/2 hours after eating; or after exercise, to wait for at least an hour before eating. Exercise immediately before going to bed should also be avoided, since stimulation of the adrenalin glands interferes with relaxation. A one-hour "unwinding" period should be allowed if exercise is performed in the late evening. Other than these considerations, when to exercise is a matter of individual preference. Some prefer early morning as a stimulation for the day or "to get it over with." Others like the noon hour to break up the day. Still others find that late afternoon or early evening offers a chance to relieve the tensions of the day. Exercising before mealtime can sometimes help those who are attempting to lose weight, since the desire for food may be less.

how often to exercise

The frequency of exercise depends somewhat on the type of program engaged in and whether the goals are for improvement or maintenance. Since deconditioning usually begins after the second day of nonexercise, a least three days a week (MWF or TTS) should be a minimum, though gains of up to 20 percent have been shown from as little as two thirty minute periods a week in personally supervised classes. Weight training programs are best conducted three days a week or every other day. Aerobic activities are recommended from four to six days a week, depending on intensity. Theoretically, calisthenic programs should be performed daily with the maintenance program done three days a week, but five days would be adequate. Combining it with an aerobics program would be ideal.

clothing

The basic rule on clothing is that it should be light, loose, and protective. For running, well-fitted, flexible rubber or ripple sole shoes with good support should be worn. Blisters may be prevented by wearing two pairs of socks, with a thin and lighter pair underneath a heavier pair to minimize rubbing and absorb friction—the primary cause of blisters. White socks are preferable to avoid possible infection from the dye of colored socks should blisters occur. Loose fitting shorts will prevent chafing of the skin. Shirts should be light and airy enough to provide for good ventilation. Rubber suits should not be worn, since they tend to raise an already high body temperature, placing added stress on the heart, and may result in excess sweating. Sweatshirts should be worn only in cold weather for similar reasons.

Sweating and dehydration. Excess sweating can lead to dehydration of tissues and loss of salt which can cause heat cramps, heat exhaustion, and heat stroke. Heat exhaustion is reflected in a weak and rapid pulse, cold skin, and dizziness. Treatment involves adequate rest and the intake of adequate fluid. Heat stroke symptoms are a hot, flushed skin (usually dry), high body temperature and possible delirium. It demands both immediate medical attention and attempts to lower body temperature and prevent shock.

Though weight loss accompanies excess sweating, the loss is temporary until fluid is taken in and of no real value. Dehydration resulting in as little as 2 percent loss of body weight has been shown to adversely affect physical performance. The only possible reason for deliberately causing excess sweating might be following an abnormal intake of fluid.

Salt tablets taken before or after exercise are often used to prevent dehydration in warm temperatures, but they often cause stomach upset and nausea. A better method is to make a more liberal use of table salt with meals, supplemented in severe conditions by adding salt to a refreshment such as lemonade at about one-half to one teaspoon for each four glasses of water.

Caring for blisters. Blisters should be cleaned with an anesthetic solution and covered with a light gauze. Taping gauze on the back of the heels or over an irritated spot on the sole of the foot may help to prevent the full formation of a blister.

Preventing athlete's foot. Clean socks and shoes will help to prevent fungus between the toes which can cause itching, burning, scaled skin, and possibly a more serious infection. When showering, care should be taken to dry thoroughly between the toes. Airing shoes and feet and sprinkling powder between the toes are other good practices.

warm-up

Before beginning any vigorous activity, the body should be given time to adjust from the resting to exercise condition. Warming up is particularly important for the prevention of muscle soreness and/or injury, since both speed of contraction and relaxation is slower when muscles are cool, with the relaxation phase relatively slower. Driving muscles into contraction before they have completed their relaxation phase can result in what is called a *summation of contractions.* This creates a hardening effect and subsequent muscle soreness. Because of decreased viscosity (resistance to internal rearrangement), muscles and tendons are also more easily stretched after being warmed up which can help prevent strain or tear.

To be most effective, the warm-up should include whole body activity that raises muscle and blood temperature enough to produce sweating, but not so strenuous as to cause undue fatigue. Wherever possible, it should be closely related (if not identical) to the activity to be performed. This is especially true for sports activities. The effects of an adequate warm-up have been shown to last from forty-five to eighty minutes.

exercise sequence

Though there is no scientific evidence to support it, the following warm-up exercise sequence has proven satisfactory: (1) whole body activity, such as jumping jacks or running-in-place; (2) trunk and hip; (3) arm and shoulder; (4) neck; (5) leg and knee; and (6) ankle and foot. Conditioning exercises can follow a similar pattern, though weight trainers usually prefer to begin with arm-shoulder-chest exercises.

Trunk movements. There is a sound physiological basis for preceding forward trunk flexion (bending) movements with trunk rotation and side flexion. The muscles of the back should be given a release from the tension of maintaining an upright position. In bending forward, the back muscles actually contract while lengthening (called eccentric contraction), in resistance to the force of gravity. This means that these muscles, though stretched, are not relieved of tension. Lateral rotation and side flexion, on the other hand, provide for an alternate contraction and relaxation of these muscles, thus helping to release tension and make forward bending easier. Another recommended practice is always to follow back hyperextension exercises, such as back arches, with back stretching exercises.

leg lifts and straight-leg sit-ups

Back lying leg lifts and straight-leg sit-ups should be avoided as abdominal exercises, especially by those with an excessive curvature of the lower spine (lordosis) and with low back problems, because of the attachments of the muscles involved in these movements. The muscles are the rectus femoris (front of the thighs) which attach to the front of the pelvis, and the psoas muscles (hip flexors) which run from the top of the lower leg to the front of the lower spine. When these movements are performed, there is a tendency to pull the pelvis forward and downward and the lower spine forward, thus aggravating the lordosis condition. If the abdominals are relatively strong, they may be reasonably successful in stabilizing the pelvis during these movements, but still not without strain on the lower back muscles. If the abdominals are weak, the pelvis may not be stabilized at all. To strengthen the abdominals, the bent-knee sit-ups are best, since the other muscles are neutralized.

arm support exercises and isometrics

For middle-aged and older persons, arm support exercises should be kept to a minimum and isometric exercises are not recommended at all. Both tend to raise blood pressure, shorten breath, and place unnecessary stress on the cardiovascular system. For those with high blood pressure or some narrowing of the arteries, this can be dangerous. For those who have continuously maintained a good level of condition, the precaution is probably not needed.

exercise cadence (rhythm)

Unless speed overload is being used or the rate of speed is a factor, the exercise cadence should be moderately slow and rhythmic. As noted, this is especially so in flexibility exercises, but since even strength and endurance exercises should involve a full range of movement, the same principle should be applied in these exercises.

sprinting

Sprinting is basically an anaerobic activity and builds up a rapid oxygen debt. It has value for building up stroke volume and for training to improve ability to tolerate an oxygen debt, such as required for certain sports activities (football, wrestling, etc.). For overall cardiovascular improvement and endurance training, however, sprinting is unnecessary. The common practice of ending a long-distance endurance training run with a sprint is also unnecessary and should always be avoided by most middle-aged persons. The sudden demand for blood and oxygen may well be beyond the reserve capacity of a heart already stressed—with possibly fatal results.

breathing

Holding the breath should be avoided during exercise. Holding the breath during muscle contraction closes the glottis (opening between the vocal cords in the larynx at the upper end of the trachea, or air passageway), which causes an increase in intrathoracic (chest) pressure. This decreases venous blood return to the heart causing a lower stroke volume, less cardiac output, and a loss of oxygen at a time when it is needed most. The condition is known as the "Valsalva" effect, and can be dangerous if coronary disease, hypertension, asthma, or emphysema is present. At the least, it may cause a temporary "blackout." Improper breathing may also cause tightness in the neck, dizziness, or headaches.

Forced expiration on each contraction will help to avoid breath holding and automatically trigger an adequate inspiration which may be accomplished during the relaxation phase of the exercise. It will also help to tense and stabilize the entire body. This is particularly important in weight training and in contact sports for added force and protection. In calisthenics, an example would be to exhale during a push-up extension and inhale while lowering; in running, exhaling every other time the left (or right) foot strikes the ground. During running or heavy exercise, a rate of breathing higher than 13 to 15 for fifteen seconds is a signal to slow the pace or intensity of the exercise (one per second).

"stitch in the side"

During the early respiratory adjustment to heavy exercise, especially running, a sharp and severe pain is often felt just below the rib cage and to the side. The cause is assumed to be a local and temporary blocking of circulation (ischema) in the muscles of the diaphram or supporting intercostal muscles. By slowing the pace and allowing time for blood diffusion to take place in the tissues, the pain will shortly disappear. A good warm-up can help to lessen the severity, if not avoid it completely. As condition improves, it occurs less frequently or not at all.

"second wind"

Another phenomenon often experienced during endurance-type activities even among well-trained athletes is an initial, very labored breathing, followed by a feeling of relief and a return to normal breathing. It is essentially the result of the body's delay in its metabolic adjustment to the exercise. Because the phenomenon has usually been observed at about the same time as sweating begins, it seems to be related to changes in muscle efficiency as the muscle temperature is increased. The importance of a good warm-up is again indicated.

chest pain or undue fatigue

Sharp chest pain or undue fatigue during exercise should always be a signal to stop and rest, especially for those over thirty or those with a history of heart disease. A medical check and clearance should be made before resuming exercise. The same precautions should be taken for chest pain or prolonged fatigue following exercise.

cooling out

Vigorous exercise should never be ended abruptly; that is, followed by immediate rest (sitting or lying down). Abrupt stopping decreases venous blood return to the heart, decreases stroke volume, and causes a rapid increase in heart rate to maintain an adequate blood flow. In effect, the heart may be left beating on an almost empty chember. At the least, dizziness or fainting may result, but shock is also a possibility. Among older persons, most heart attacks occur during this period.

To maintain an adequate venous return and allow time for the body to readjust from the exercise state, slow jogging or walking about while swinging the arms should be continued until the pulse rate has dropped to at least 120 (a ten-second count of 20). This will also help to relieve muscle tightness caused by prolonged running.

postexercise heart rate

A two-minute postexercise heart rate check (two minutes after stopping exercise) is a good index of the heart's tolerance to exercise and the intensity of the exercise load. Regardless of age or level of condition, if the heart rate is above 120, it is a sign that the exercise load was too intense and should be reduced. For women, of course, a higher allowance should be made (6 beats per minute, or 21 for a ten-second count). If the count is above 120 five minutes after stopping, a medical examination would be indicated.

showers

Hot showers and saunas. Hot showers, saunas, or steam baths should be avoided after exercise, since both tend to elevate heart rate and place additional stress on a cardiovascular system that has already been worked enough. Serious accidents and even fatalities have occured as a result of these practices. As a method of relaxation these practices are fine, provided they are not combined with a heavy exercise program at the same time.

Ice-cold showers. Ice-cold showers over the chest have been shown to increase blood pressure, heart rate, and cardiac output. For healthy individuals, this is not dangerous in itself, but it could be dangerous to those with cardiovascular problems—known or unknown. A moderate temperature (70 degree) shower following exercise should be the basic rule.

environmental conditions

Heat and especially the combination of heat and humidity place added stress on the heart. Higher pulse rates and blood pressures are to be expected. The exercise intensity during these conditions should be reduced, along with the warm-up time. Extremely cold temperatures also have the same effect and similar precautions should be followed. It is during cold weather that sweatsuits are advisable. A longer warm-up is also warranted.

muscle soreness, shin splints, and cramps

We have already seen that muscle soreness at the beginning of a training program can be expected, but a good warm-up which includes stretching exercises can help to relieve it. In any event, as the exercise is continued and blood circulation improves, the soreness will disappear.

Painful shin splints (pain along the sides of the shin bones caused by a slight tearing of the connective tissue next to the shin bone) and excessive calf soreness can be avoided in running programs by not running on hard surfaces, such as street pavements. The running should be done on grass or dirt instead. It is also important that a proper running technique be used.

Running technique. According to Bill Bowerman, outstanding track coach at the University of Oregon and a man who has conducted innumerable tests and experiments with runners for over twenty years, the most important element for a smooth and efficient running style is maintaining an upright position.[7] It not only conserves more energy, but allows the greatest freedom, ease of movement, and mechanical advantage. This is just as true, he says, for sprinters as it is for long-distance runners or joggers, unless one is engaged in a contact sport, such as football, where a forward body lean is needed for greater force upon impact. Except for slight modifications in the arm, leg, and foot-strike actions, this means that the best running position is essentially the same as for walking.

The head should be kept up, the back naturally and comfortably straight, and the body line from the head through the hips perpendicular or nearly so. A very slight body lean forward is permissible, but a rigid military posture with the shoulders thrown back should definitely be avoided. Thrusting the shoulders back costs more energy, causes quicker

[7] Bill Bowerman, "The Secrets of Speed," *Sports Illustrated*, 35, No. 5 (August 1971), 22–29.

fatigue, tends to cause an ache between the shoulder blades, and puts more strain on the lower back. To avoid dropping the head and putting strain on the neck, the eyes should look straight ahead rather than at the feet or ground, with the head held steady.

The arms should be bent, with the elbows held slightly away from the body and above waist level. They should swing in a slightly diagonal plane and in a natural counterbalancing motion to the movement of the legs and length of the stride. The legs should swing freely from the hips with a knee lift leading the way, the ankles relaxed, and the whole of each foot contacting the ground and rolling up with each completed step. Either the heel-to-toe or flat-footed footstrike can be used.

In the heel-to-toe technique, care should be taken not to land on the base of the heel, but on the mid-point or fuller part. After landing, rock forward to the ball of the foot and take off for the next step. Hitting with the heel first cushions the landing and the rock forward distributes the pressure. The foot should strike the ground after it has reached its furthest point of advance and has started to swing back. As it becomes flush with the ground, the knee should be in a direct line with the foot, providing maximum support for the body. This technique is the most natural way to run, is least tiring, and is used by the great majority of all good long-distance runners.

In the flat-footed technique, the whole foot strikes the ground at the same time, directly in line with the knee, in a quick light-stepping action. Care must be taken not to drive down hard on the foot, but to let it pass under the body and quickly pick it up for the next step just as contact is made.

For short distance sprinting, knee lift is increased, arm thrust is accentuated, and foot contact made on the ball of the foot to provide more driving force and eliminate unnecessary foot action. When used for long-distance running, however, the ball-of-the-foot technique is awkward, uncomfortable, and harder on the calves and shins.

If shin splints do occur, the soreness can be relieved by a series of foot flexion, extension, and rotation movements while in a sitting position. (See Exercise 12, p. 109.) Sometimes it may be necessary to stop running for a few days until the pain lets up. Should a muscle cramp occur, the best procedure is to immediately stretch the muscle.

exercise and alcohol

Alcohol and exercise simply don't mix. Alcohol interferes with the enzyme action which assists the transfer of oxygen from the hemoglobin in the red corpuscles to the working tissues. The decreased oxygen supply is the reason that alcohol dulls the senses and fatigues the body.

In addition, alcohol causes a constriction of the blood vessels in the heart. At a time when oxygen is needed most, the resulting oxygen deficiency during vigorous exercise could prove extremely dangerous. The better conditioned, of course, the less the risk. As a general rule, alcohol should not be taken for at least four hours before exercising.

exercise and smoking

That smoking "cuts wind" and prevents the achievement of top flight condition has long been recognized by athletic coaches, even if based only on empirical evidence. Recent research has provided scientific evidence which supports the customary ban on smoking for members of athletic teams and which should be considered by anyone interested in raising his level of fitness through exercise.

In a study reported in the *Journal of Applied Physiology*, it was shown that fifteen puffs of cigarette smoke in five minutes caused an average increase in airway resistance of 31 percent in the resting state. Changes occured as early as one minute after smoking began and lasted from ten to eighty minutes.[8] The effect on the efficiency of breathing under exercise conditions when maximum ventilation is needed can easily be seen. In addition to increasing airway resistance, the cigarette smoke increases the amount of carbon dioxide and carbon monoxide in the oxygen-carrying hemoglobin blood cells, thus lowering the amount of oxygen in the arteries and poisoning the ability of the blood to deliver oxygen to the working tissues. A decrease in heart volume or size has also been noted as an effect of cigarette smoking, though whether from an actual decrease in muscle size or from a decreased blood supply is not known. Significantly, not one smoker we have tested has been able to make a full Heart Recovery Rate on the step test explained in Chapter 2—even those who exercise regularly. Those who made a full recovery were invariably nonsmokers. These facts make it evident that the smoker can never reach the high levels of fitness possible for the nonsmoker.

exercise and diet

Many people needlessly spend money on food supplements (extra vitamins, protein, minerals) advertised as being able to improve physical performance or speed up muscular development. There is no sci-

---

[8] J. A. Nadel and J. H. Comroe Jr., "Acute Effects of Inhalation of Cigarette Smoke on Airway Conductance," *Journal of Applied Physiology*, 16 (1961), 713–16.

entific evidence available that supplementing a basically sound diet can accomplish such results. If the basic diet is inadequate, food supplements can be of value, though in the absence of medical advice, reliance on natural sources would appear to be the better common sense approach.

The body can derive its energy from food carbohydrates, fat, or protein, but in muscular activity, it seems to prefer burning carbohydrates. Though fat can provide twice as much energy per gram as carbohydrate, the oxygen cost for breakdown is higher. High fat diets have been shown to bring about earlier fatigue and are, therefore, not recommended for endurance-type activities. Protein is used primarily for tissue building and is drawn upon for energy only under starvation conditions. For muscle-building activities or other strenuous, high energy activities where muscle mass is likely to increase, an increase in protein with a proportional decrease in fat consumption is recommended.

A word of caution should be given about combining a dieting and exercise program for losing weight. A very low calorie intake may cause problems caused by low blood sugar and a depletion of glycogen or energy stores in the liver. When followed by exercise, this condition can be dangerous, since muscle contraction may be forced to stop. If it occurs, the victim should be given high energy food such as sugar immediately.

It is best that a severe dieting program precede an exercise program. If both are done concurrently, a wise precaution would be to increase carbohydrate storage just before the workout by a high carbohydrate snack (cereal and milk, toast and honey, etc.). There is evidence that such a meal taken up to thirty minutes before a one-mile competitive run has no adverse effects. Chapter 7 goes into detail on a sound procedure for controlling weight through exercise and diet.

special exercise machines

On the basis of what has been said about the need for overload and adaptation to improve motor and cardiovascular fitness and what must be done to make changes in other areas, it should be evident that there are no shortcuts to fitness. Improvement requires time and effort directed specifically at all the components of fitness. Despite advertised claims, there is no evidence that special vibrating, massage, or other machines which passively exercise the body are of any significant value in raising a person's fitness level. As a surface stimulation to circulation, for relaxation, or for the loosening up of thick, heavy, fat deposits, these devices can be of some value. Otherwise, you'd do better to save your money.

figure control studios

There are many fine commercial organizations which specialize in body building or figure control for men and women through programs employing weight training equipment, various exercise machines, saunas, and wet baths. Many are staffed by trained personnel and conducted in accordance with sound physiological principles. Others, however, must be viewed with great skepticism, particularly those which make exaggerated claims of being able to give anyone an "ideal" build, figure, or dress size through "revolutionary new machines or methods." Often people are lured into signing expensive contracts after initial advertising which suggests "fantastic results at a cost of just $1 a week" or less, with relatively no effort and losses of "up to 8 inches from the waist, hips, and thighs on the first visit."

In California and, perhaps, elsewhere, several civil suits were filed against such firms for false and misleading advertising and deterioration of health as a result of the prescribed diets and exercises. The plaintiffs said that they were deprived of food and nutrients necessary to health and became excessively tired, strained, and exhausted, resulting in pain and health problems forcing medical attention. They disclaimed any advertised "incredible new method" giving the results claimed and indicated that part of the "new" program was the use of rubber suits which was included only as part of a "full contract"—costing $496. A stipulated judgment and civil penalty were given against a particular studio which was also enjoined against advertising that they have made weight reduction a science and against offering any guarantee unless described in detail.

With a basic understanding of the principles, techniques, and precautions given in this chapter, you should be in a better position to judge the worth of such "instant cure" solutions to your fitness problems. Exercise for improvement and maintenance (as all other methods) is a long-range process that cannot be hurried or rushed. It demands regular participation and should be viewed as relating to lifetime goals. You simply can't expect dramatic results in just a few days or weeks. If you're "out of shape," remember it took time for you to get that way and it will take time to get back in condition. With patience, and by applying in a variety of sound exercise programs the principles, techniques, and precautions presented, you'll eventually achieve your fitness goals with more safety and at far less expense.

individual
exercise programs

The purpose of this chapter is to present and analyze some of the various exercise programs and activities that can be used to develop and maintain a good level of fitness. There are many sound programs that can be followed or developed to meet individual needs and preferences. The activities discussed here have been selected on the bases of proven success, incorporation of sound principles, and adaptability to an individual exercise program. It would be a good idea to become familiar with all of them, and then choose the ones that are most specific to your particular needs. As your needs and opportunities change, it will provide you with the adaptability to change with them.

The programs, of course, can be performed with another person or in a group situation (some need the motivation of others working with them), but remember that for lasting benefits any exercise program must be continuous. Ultimately, this means being self-directed and self-motivated without reliance on others. It will take time for changes to take place, but once you experience them, it will provide your greatest motivation to continue.

Before beginning any of the programs, we suggest you compute your minimum and maximum Exercise Heart Rate Range and record your present "normal" sitting and standing Resting Heart Rate, so that they can be used as both a guide and basis for measuring improvement. A page is provided at the end of the chapter for these computations, recordings, and future adjustments.

We have already discussed the importance of an adequate warm-up. It should begin gently and gradually progress to more vigorous action as you become prepared for the full workout. The exercises presented here are applicable to any conditioning program or sports activity. If done prior to a sports activity, they should be followed by moderate execution of the various sports skills and related activities to be performed. The exercises place an emphasis on various stretching movements and can be used also as a daily exercise program for improving flexibility.

A sound approach is to use the recommended number of repetitions in moderate weather, increase the number in cold weather, and reduce them as the temperature rises. Do the exercises in the order listed. In all stretching movements, try to gradually increase the degree of stretch with each repetition, but be sure to do them slowly and gently, not abruptly. Also, record your progress on the Warm-Up and Calisthenic Exercise Guide on page 139.

1. *Jumping Jack* (two counts): Stand with arms at sides. Jump, spread the feet to the side, and simultaneously swing the arms overhead on count 1; then swing the arms down and jump back to the starting position on count 2. Use a rhythmical and moderate cadence (15 reps.).

2. *Side Bender*: Stand, feet together, one arm extended straight upward and the other at the side. Slowly bend to the side of the down arm as far as you can go, bob gently (start a short return movement, but bend and stretch again), then come up, switch arm positions and repeat to the other side (5 reps. each side).

3. *Side Twister*: Stand, feet comfortably apart, with arms extended out to the sides, palms down. Slowly twist to the side as far as you can go, bob gently once, and repeat to the other side. To relieve strain on the knee, turn the far foot slightly in as you twist (5 reps. each side).

4. *Forward Bend and Stretch*: Stand with feet shoulder width apart, with the hands hanging loosely at the sides or placed on the hips. Bend the knees slightly, curl the head and trunk, extend the arms between the legs, and gently reach to touch the ground at about heel level. Bob once, trying to reach further back, and then come to a standing position, thrusting the shoulders backward (5 reps.).

5. *Elbow Thrust*: Stand with feet comfortably apart, with the arms bent, hands in front of the chest, and the elbows out to the side. With-

out arching the back, rhythmically thrust the elbows backwards, and return to the starting position (15 reps.).

6. *Shoulder Roll*: Stand with feet comfortably apart, with the fingers touching the shoulders. Slowly rotate the elbows in a full circle forward, up, back, and down for 5 repetitions. Reverse direction and repeat for 5 more.

7. *Neck Roll*: Stand with feet comfortably apart, with hands on the hips. Gently roll the head in a full circle first to one side, then to the other (3 reps. each side).

8. *Knee Lifts*: Stand with feet comfortably apart, arms at the sides. Raise one knee to the chest, grasp and pull it to the chest; return and repeat with the other leg. Keep the back straight (10 reps. each side).

9. *Knee Bends*: Stand with feet comfortably apart, hands on hips. Bend the legs to just short of a 90 degree angle, extending the arms forward for balance as you go down, then return and repeat in a slow to moderate cadence (10 reps.).

10. *Toe Touch*: Stand with feet together, with hands on the hips. Keeping the knees locked, bend and gently reach towards the toes as far as possible, attempting to touch the toes or floor. Return and repeat in a slow rhythm (10 reps.).

11. *Toe Raises*: Stand with feet comfortably apart, with hands on the hips. Raise to your tiptoes, extending upward as far as possible, and return. Repeat in a slow rhythm (10 reps.).

12. *Ankle Roll*: Sit with the legs spread, hands on the floor at the buttocks, and leaning back slightly. Slowly rotate the ankles in a full circle, first in one direction and then in the other (10 reps. each way). This may be done standing by placing the hands on the hips, and then rocking forward, out to one side, back on the heels, and to the other side, reversing direction after 10 repetitions.

13. *Slow Jog*: Stand in place with the arms in a running position. Slowly jog in place or in a small circle for 50 counts, counting each time the left foot strikes the ground. Begin slowly and pick up the pace gradually after every 10 counts, until the last 10 are at full speed with the knees high.

14. *Deep Breathing*: Stand with feet comfortably apart. Slowly swing the arms forward and upward, raise up on the toes, and inhale deeply until the arms are in an overhead position. Swing the arms down, drop to the heels, and exhale as the arms are returned to the starting position (5 reps.).

general calisthenics program

A complete calisthenics program includes a variety of exercises aimed at developing flexibility, coordination, minimal muscular strength and endurance, muscle tone, and body lines. It should never be accepted as a complete program, however, since calisthenics are only of limited benefit to the heart, lungs, and vascular system. The warm-up exercises, followed by the series of conditioning exercises listed below, and one of the aerobics programs to be described, would be an excellent combination for developing all the physical components of fitness. Following the warm-up, do the exercises in the order shown. Depending on your present condition, start somewhere between the minimum and maximum number of repetitions shown and gradually increase to 'the maximum number indicated for maintenance. Speed overload may be used to increase the overload. If desired, the calisthenics program may be done following the warm-up and aerobics activity.

top to bottom program

1. *Toe Touch Series*: Four standing positions for this exercise, starting with the feet at shoulder width, then together, and finally crossing one foot over the other and then reversing positions. In rhythm, gently reach and touch the toes and return to the starting position (2 counts: 3–6 repetitions each position).

2. *Knee Bends:* Same as warm-up exercises (10–20 reps.).

3. *Sprinters' Drive*: Place hands on the floor at shoulder width and lean forward with one leg well up under the chest and the other fully extended to the rear. Alternate leg positions in a two-count rhythm: (1) shift positions; (2) return to starting position (5–20 reps.).

4. *Push-Ups (men)*: Lie on floor with hands directly under the shoulder joints, fingers pointing straight ahead. Extend the arms and raise the body in a straight line from head to heels to a fully extended position supported by the arms and toes. Lower the body in a straight line by bending the arms until the chest just touches or comes within an inch of the floor. Repeat in a moderate rhythm (5–20 reps.). *Women*: Same, except keep the knees in contact with the floor throughout the movement. *Note*: Older persons may begin with the hand-knee push-up.

5. *Sitting Stretch-Static*: Sit with feet together, hands at the sides. Without bending the legs, bend the trunk forward, tuck the head, reach forward as far as possible, and grasp firmly around the legs, ankles, or feet, according to the extent of your reach. Hold for six seconds, relax, and return to the starting position (5–10 reps.).

6. *Sitting Stretch-Ballistic*: Sit with the legs spread and straight. In four counts: (1) bend forward, extend the arms, and touch one foot; (2) come back slightly and reach out between the feet; (3) come back slightly and touch the other foot; and (4) return to the starting position. Do in a gentle and rhythmic manner (5–10 reps. each direction).

7. *Gather Sit-Ups*: Back lying position with legs together and ex-

tended, and the arms extending overhead. In one motion, curl up, bring the knees up, and wrap the arms around the knees at the middle. Return to the starting position and repeat in a moderate rhythm. In raising up, roll the head, neck, shoulders, upper and lower back in order and then unfold in the reverse order coming down (10–30 reps.).

8. *Leg Cross-Over* (*sitting*): Sit with the legs extended, feet together, hands on the floor by the buttocks and leaning slightly backward. Keeping the hips in place, lift one leg, bring it across, and touch the toe to the floor as high up as possible. Return to the starting position and repeat to the other side (5–20 reps.).

9. *Side Leg Lifts*: Lie on one side with the legs together, the head supported by the elbow and hand, and the other hand on the floor in front of the body for balance. With the leg straight, lift it as far as possible and return to the starting position. Repeat in a moderate cadence (5–20 reps. each side).

10. *Back Leg Lift*: Front kneeling position, with hands on floor at shoulder width. In two counts: (1) Bring one knee to the chest and at the same time tuck the head, attempting to touch the nose to the knee; (2) Extend the leg back and up as far as possible, with the knee slightly flexed and at the same time lift the head and shoulders as far as possible. Repeat in a rhythmic cadence (5–20 reps.), and then repeat the procedure with the other leg.

11. *Trunk Raise*: Front lying position, hands under the thighs, legs straight and together. Keeping the legs in contact with the floor, simultaneously raise the head, throw the shoulders back, and lift the trunk as high as possible. Return and repeat in a moderate cadence (5–20 reps.). (Skip if you have low back problems.).

12. *Flutter Kick*: From the same starting position, this time keep the chin and trunk in contact with the floor and lift the legs until you feel the thighs begin to clear the fingers. Keeping the legs straight, alternately move the legs up and down as in a swimming movement in a moderately fast cadence. Count each time left foot descends and repeat for from 10 to 40 counts. (Skip if you have low back problems.) End by pushing up, sitting back, and tucking head forward (5 reps.).

13. *Jog and Run-in-Place*: As a windup, take the running position and do an alternate slow jog and run-in-place at maximum speed in intervals of 10 counts each (6–12 intervals).

weight training

Weight training gained popularity through the use of progressive resistance exercises in rehabilitation hospitals during and following World War II. Since then there has been much research verifying its effectiveness not only for rehabilitation, but for correcting postural defects, increasing athletic performance, and bringing about desirable changes in body appearance.

Weight training is not to be confused with weight lifting. Weight lifting is a competitive sport concerned with how much weight an individual is able to lift overhead using certain types of lifts. Competition is by body weight divisions, similar to that of boxing or wrestling. It is a sport that holds local, national, and international competitions, including the Olympic games. Weight training is not concerned with competition, but with the development of increased physical capacity, especially muscle strength, endurance, power, and speed of movement. When conducted properly, flexibility can also be improved.

advantages and disadvantages

The advantages of weight training in a fitness program are: (1) exercises can be selected to develop specific areas or muscles of the body; (2) a variety of exercises are available for relief of monotony and boredom; (3) the exercises are simple in nature and involve skills that can be quickly learned and practiced; and (4) although there are highly satisfactory machines available in schools and clubs, a minimum of equipment is required that is inexpensive, handy in nature, and easily stored.

From a total conditioning point of view, the disadvantages are that it is not an all-inclusive program. It does not: (1) develop all motor fitness components (agility, coordination, balance, dynamic flexibility); (2) make a strong enough contribution to cardiovascular fitness; and (3) develop efficiency in motor skills.

equipment

Private clubs, public gymnasiums, and schools or colleges often have expensive machines that can be used most efficiently for both group and individual workouts. An individual commercial set of equipment, however, can be purchased at reasonable prices. For a personal home weight training program, the following basic equipment is needed: a bar, plates, collars, and dumbbells.

Bar. This is a steel bar usually 4 to 6 feet in length, 1¼ inches in diameter, and weighing about 20 pounds. It is used to support the collars and plates. A knurled middle area provides for firm gripping by the hands to prevent slipping, loss of control, and possible injury.

Plates. The plates are round in shape and sized according to weight, with a hole in the middle to slide them on and off the bar. The plates or "weights" come in sizes ranging from 1¼ pounds to 100 pounds.

Collars. Collars are the protective locks used to secure the plates on the bar. Set screws are generally used to attach the collars, with one set of collars on the inside and one set located outside the plates. The proper setting and securing of the collars is important to prevent the plates from sliding and upsetting the balance of the total lifting weight. The bar and collars together usually weigh about 25 pounds and are included with the plates in determining the total poundage.

Barbell. Barbell is the term used to describe the completed setup when ready for an exercise, including the bar, selected plates, and collars.

Dumbbell. Originally, the dumbbell was an end-weighted piece of equipment cast in one section, about 12 to 18 inches long and of varying poundages. Though still available in solid construction, small bars with changeable collars and plates are now more commonly used to allow for greater adaptability. They are primarily used with one-arm lifts, while the barbell is used for two-arm lifts.

basic terms

As most other activities, weight training has its own unique language. Knowing the language will help in understanding directions and learning the rules and techniques of the activity.

Repetitions. Number of times a complete exercise movement is performed.

Sets. Number of times a given number of repetitions are performed, usually consecutively after a brief rest.

Load. The actual weight or poundage used for an exercise or in a set, including the bar, plates, and collars.

Repetitions maximum (RM). The maximum load that can be lifted a given number of times. Thus, 10 RM is the greatest weight that can be lifted 10 times. (Sometimes also called Execution Maximum—EMs.)

Minimum or maximum repetitions. The minimum or maximum number of repetitions performed with a given load.

program principles and procedures

There are many theories and practices regarding weight training, but only a few formal studies to substantiate them. There are, however, a number of practices and acceptable theories that have evolved and, with the studies, can be used to give guidance to the beginning weight trainer.

Balanced exercises for symmetrical development. A well-rounded weight training program should work all major muscle groups of the body equally in order to insure balanced development.

Objectives of the program. Though all programs must include a balanced set of exercises, the procedure to be followed in terms of load, repetitions, and sets will depend on the objectives of the program: strength, muscle endurance, combination strength and muscle endurance, or hypertrophy. All these elements will be improved with any procedure, but some routines are more effective than others when development of a particular element is desired.

Single and double methods of progression. The single method of progression involves either keeping load constant and gradually increasing repetitions (when muscle endurance is the primary concern), or keeping repetitions constant and gradually increasing the resistance (when strength development is primary). The double method involves a gradual increase in repetitions with a given load from a minimum to a maximum number, then adding resistance and repeating the procedure. In both methods, the progressive increase in load or repetitions is done on a weekly basis.

Strength development. Most studies show that strength gain is best developed by doing two or three sets of from 4 to 10 RMs. Three sets of 6 RM each appears to be optimum.

Muscular endurance. Muscular endurance appears to be best developed when the load is kept high enough to constitute an intensity overload and develop strength as well. In other words, merely increasing repetitions with a very light resistance, even if up to 100 times or more, will not do the job. For emphasis on muscular endurance, increasing repetitions to between 15 and 25 RMs seems to be most effective. The only advantage of repetitions beyond this is muscular definition (outline). In summary, then, it can be said that "strength is a prerequisite to endurance," and a certain minimal level must be achieved before emphasis can be placed on training for muscular endurance.

Combined training for strength and endurance. One method of combining strength and endurance training is a program using 10 repetitions maximum. Another excellent way is to use the double-progressive method; beginning with 5 RMs, gradually increasing to 10, then dropping back to 5, adding to the load and repeating the procedure. For leg and waist exercises, a 10 to 20 RM procedure can be used, increasing repetitions by 2 until the maximum is reached. By increasing the number of sets to two or three, there can be even greater gains.

Hypertrophy (bulk). For an increase in muscle size, but not necessarily a corresponding increase in strength and endurance, a procedure that has been proven effective is to perform three consecutive sets of 10 repetitions each using loads of 1/2 10 RM, 3/4 10 RM, and full 10 RM respectively.

Power. Once a desired number of repetitions is reached, resistance can be increased with improvement in both strength and power by speed overload—performing the exercises at a faster rate.

Loads. A weight trainer should lift only the load he is able to control through a complete range of motion and according to the prescribed exercise form for the required number of minimum repetitions called for.

Starting loads. During the first week of training, light loads should be used in order to develop the correct form for each exercise and prepare the muscles for the gradual application of overload resistance. After the pretraining preparation, use the recommended starting loads shown in the basic program description in Table 4. They are based on percentage of body weight, but may be adapted upward or downward slightly as necessary. Use as your guide your ability to control the load as described above.

Form. Correct form is essential both to insure symmetrical development and to protect against injury. Each exercise is designed to work a specific muscle or muscle group. When done incorrectly, other muscles are brought into play and the intended effect of the exercise is lessened. There is also the danger of using weaker muscles to do a job they are not equipped to do, resulting in muscle strain. If you can't make a lift while keeping correct form, it means the load is too heavy for you and should be decreased.

Range of motion. An important aspect of correct form is performing each exercise movement through the complete range of motion at the joint involved. Failure to do so will result in only partial development of the primary movers and a loss of flexibility in the antagonist muscles. Failure to include a balance set of exercises, using improper form, and not working the muscles through their full range of motion may all result in developing a muscle-bound condition.

Bound muscle. Lost range of motion at a joint caused by over-developed or improperly developed muscles is what is meant by "muscle-bound." Obviously, a shorter range of movement can occur even when all muscles are developed symmetrically. Greater bulk in the forearm and biceps, for example, brings them closer together in any flexing movement and limits the extent that the triceps can be stretched. The more common cause of muscle boundness, however, is unbalanced development of the antagonistic muscles at a joint, or the overdevelopment of one muscle at the neglect of its antagonist.

Cadence or rhythm. Unless speed overload is being used to develop power, most weight training exercises should be performed at a moderate cadence. This helps to preserve proper form and full range of motion.

Breathing. To insure against dizziness or blackout from the Valsalva effect, try to breath freely and naturally and avoid breath holding. The basic rule is to exhale during exertion and inhale during the relaxation phase. An exception may be made when using relatively light loads and when the exercise creates a natural expansion of the rib cage during exertion, such as Lateral Raises—lying and bending. (See the program.) For greater stability when lifting a heavy load to a starting position at the chest, it is also better to take a deep breath just before the short lifting action is made and hold it until completed for greater stability.

Workout frequency. Generally, workouts three times a week with a day of rest in between or workouts every other day are best. Working out daily can fatigue the muscles unless the exercise routine is alternated between upper and lower body exercises.

Workout time. Workout time will vary according to the type of program used, the time available, and the experience of the individual. For a beginner, a thirty- to sixty-minute workout will be sufficient. It can be lengthened as strength, endurance, and experience are gained.

Warm-up. Warm-up should always precede the regular workout. This is best done by using submaximal or light poundage exercises to lightly work the major muscle groups, but the warm-up exercises previously described can also be used.

fundamentals and techniques

Grips. There are three basic barbell grips: the *overhand grip,* the *underhand grip,* and the *combination grip* (one hand over and the other under). The grip used will vary with each exercise, and whether the palms face up or down will depend on the starting position. The overhand grip is the one most used and is sometimes referred to as the *regular grip.* The underhand grip, also called the *reverse grip,* is used in only a few exercises and only when the starting barbell position is below the waist. The combination grip is also used only for a few lifts —mostly low lift exercises in order to provide greater control.

Hand spacing. The distance between the hands when taking a grip depends on the exercise performed and may be either: (1) close (hands touching); (2) normal (about shoulder width apart); or (3) wide (out to the collars).

Starting positions. Starting body positions may be either standing, supine (back lying), sitting, or bent-over (trunk inclined forward), with the legs either bent or straight. The barbell may be either across the thighs, chest, or shoulders (also referred to as the front-hanging, chest-rest, or shoulder-rest positions). In the front-hanging position, the arms are fully extended and the bar may be held with any grip. In the chest-rest position, the arms are flexed, with the forearms under the bar and the hands in an overhand grip position. The arms and hands are in this same position in the shoulder-rest position, except that the bar lies behind the neck and on the shoulders and upper back.

Lifting to the starting position. In lifting a barbell to the thigh, chest, or shoulder-rest position, a definite technique must be used to avoid injury. For lifting to the thigh position, the feet should be spaced about shoulder width and placed well up under the bar, with the bar close to the ankles. The body should then be bent at the knees and the hips at about a 45 degree angle with the head and back straight and the arms reaching for the bar. It is then important to set the proper grip for the exercise (close, normal, or wide) and check to see that both hands are equal distance from the collar on their side to insure that the load is balanced. After taking the grip, the head is then extended slightly and, with the back kept straight, the bar is brought to the thigh simply by straightening the legs. The arms must be thought of as mere attachments and lifting with the back avoided. In lifting to the chest position, the procedure for approaching the bar and setting the grip is exactly the same. But now after setting the grip, extending the head and making sure that the back is straight, the barbell is brought directly upward from the floor in an explosive movement to a position across the chest (the degree of explosiveness depending on the load). The initial force for the lift comes from extending the legs and pulling up with the arms. As the elbows approach hip level, they are thrust forward under the bar to a position in front of it; at the same time a slight dip is taken with the legs. To place the weight behind the neck, the bar is always brought to the chest position first. Then, it is pushed directly upward with the arms, assisted by a slight upward thrust of the legs following a slight dip, if necessary. The bar is then lowered to the shoulders as the head and shoulders are moved slightly forward.

In returning the bar to the floor from the shoulder position, the procedure is reversed. It is first pushed directly upward and then lowered to the chest. From here, the elbows are brought up behind the bar, and the hands are lowered to the thighs while keeping the head and back straight. The barbell is then lowered to the floor by bending the legs and hips.

safety hints

1. Progressive weight training is an individual activity. No competition is involved. The important concern is the maximum development of each individual according to his capacities. Don't try to lift more than you can handle with good form. The result can be harmful. The amount you lift is not important. It is how well you perform each exercise.
2. Before beginning any lift, check to see that the collars and plates are secured tightly. Never lift a barbell or dumbbell without collars.
3. Always check to see that the plates are balanced on the barbell or dumbbell.
4. Keep your hands free from perspiration when lifting. Dry them with a towel or chalk if available.
5. When performing a lift above your head or where balance may be lost, always have someone (preferably two persons) near you to act as a spotter. If you do begin to fall, his first concern should be to grab the barbell, not you.
6. Always keep your eyes forward, your head up, and your back straight when lifting. Where your eyes go, your head and back will usually follow.
7. Always warm up before lifting.

courtesy

1. Don't distract or interfere with a person who is lifting.
2. Be willing to assist as a spotter whenever a person is in need of one. Don't wait for him to ask.
3. Replace all weights in position for the next person's use when you are through using them.

a basic program

The program shown in Table 4 is a basic one designed to symmetrically develop all the major muscle groups, and is a sound beginning program. As you develop experience, you can expand to other exercises and programs.

Though other muscle groups are also involved, the exercises are listed under the primary body area they are aimed at developing. You can determine the specific muscles worked in each exercise by the muscle identification procedure explained previously.

Recommended starting loads are indicated by figures representing a percentage of body weight, but they are only guides. Be sure you are able to start with a load you can handle the minimum number of times indicated in the double progression method shown. When you reach

table 4

a basic weight training program

Exercise	Starting Load	Reps
Chest:		
Bench Press	1/2	5–10
Straight Arm Pull-Over	1/6	5–10
Lateral Raises, lying		
(dumbbells)	10 lb. ea.	10–20
Arms–Shoulders:		
Standing Press	1/4	5–10
Biceps Curl	1/5	5–10
Triceps Curl	1/5	5–10
Back:		
Bent-Over Rowing	1/4	5–10
Bent-Over Lateral Raises		
(dumbbells)	10 lb. ea.	10–20
Forward Bending	1/5	5–10
Waist:		
Bent-Knee Sit-Ups		
(dumbbell or plate		
behind neck)	5 lbs.	10–20
Trunk Rotation	1/4	10/10–20/20
Side Bends		
(with dumbbell or		
plate in hand)	20 lb.	10/10–20/20
Legs:		
Half-Squat (less than		
90 degrees)	1/2	10–20
Toe Raises	1/2	10–20
Bent-Knee Dead Lift	1/2	10–20

A workout record card would be helpful to show progress. In a date column by each exercise, record your performance as follows:

$$\frac{\text{Sets–Reps}}{\text{load}}$$

Example: $\frac{1\text{–}5}{90}$ would mean you performed one 5 rep. set with a 90 lb. load.

your maximum number of repetitions, add to the load, drop down to the minimum number and progress upward again. Resistance increases should be from 2 1/2 to 10 lbs., depending on the size of the muscle group involved and the type of movement. If you can, try to work up to performing three sets in one hour. Do the exercises in the order shown.

exercise descriptions

1. *Bench Press: Starting Position:* Back lying with the bar across the chest, using a regular grip with normal spacing or slightly wider than shoulder width. *Movement:* Push the barbell upward to a position just above the nose with the elbows locked and return to the starting position. Take a deep breath just before pushing and exhale through the mouth as the bar is raised. Inhale deeply as it returns.

2. *Straight-Arm Pull-Over: Starting Position:* Back lying on the floor with the arms fully extended backward and grasping the bar out by the collars (wide grip) with a regular grip (palms facing up in this position). To relieve strain on the lower back, bend the legs so that the feet are resting flat on the floor. *Movement:* Bring the bar in an arc directly over the chest and continue until the bar touches the thighs, then return in an arc to the starting position. Inhale deeply just before beginning the move and exhale as the bar is brought forward and down. Inhale deeply as it is returned.

3. *Lateral Raises, lying (with dumbbells): Starting Position:* Back lying on a bench with arms extended straight out to the side, grasping the dumbbells with the palms up. Let the arms stretch down as far as possible. *Movement:* Bring the dumbbells to a position directly above the chest, continue and cross the arms as far as they will go, and return to the starting position. Alternate the arm cross with each repetition. Exhale as the dumbbells are raised and brought across and inhale as they are lowered to the side.

4. *Standing Press: Starting Position:* Standing chest-rest position with feet spread shoulder width apart, overhand grip, normal spacing. *Movement:* Push the bar upward until the elbows are fully locked, and the barbell is directly overhead, and return to the starting position. Exhale as the bar is pushed up and inhale as it is lowered. Keep the eyes pointed straight ahead.

5. *Biceps Curl: Starting Position:* Standing, barbell at thigh position, underhand grip, normal spacing, elbows close to the sides. *Movement:* Keeping the elbows in place, bend the arms and bring the bar to the chest, then lower to the starting position. On the return be sure the arms are fully extended and the elbows locked before beginning the next lift. Exhale as the bar is raised, inhale as it is lowered.

6. *Triceps Curl: Standing Position:* Standing with feet comfortably

apart, bar at chest position, hands at shoulder width apart. Push the bar to an overhead position. *Movement:* Keeping the elbows in close, lower the bar backward until the elbows are pointed straight ahead and return to the starting position. Exhale as the bar is brought up, inhale as it is lowered.

7. *Bent-Over Rowing: Starting Position:* Forward bend position, with the legs bent, trunk bent at almost a 90 degree angle, head and back straight, with arms hanging and hands in a wide overhand grip (out by the collars). *Movement:* Keeping the head and back straight and without changing the incline, bring the bar to the chest and return to the starting position. On this exercise, it is better to inhale deeply as the bar is raised to take advantage of the natural chest expansion, and exhale as it is lowered. If the load is a heavy one, however, use the basic rule.

8. *Bent-Over Lateral Raises (with dumbbells): Starting Position:* Forward bend position, with the legs bent slightly, trunk bent at a 90 degree angle, head and back in a straight line, and the arms hanging and grasping the dumbbells with the knuckles facing out. *Movement:* Inhale deeply and bring the dumbbells straight outward and upward at the shoulders as far as they will go. Return to the starting position. Inhale as the dumbbells are raised, and exhale as they are lowered.

9. *Forward Bending: Starting Position:* Stand erect, feet comfortably spread, bar at the shoulder-rest position, overhand grip, wide spacing. *Movement:* Keeping the legs bent slightly, bend forward at the waist to a 90 degree angle, keeping the head and back straight, and return to the starting position. Exhale as the bar is lowered, inhale as it is raised.

10. *Bent-Knee Sit-Ups (with dumbbell): Starting Position:* Back lying, holding the dumbbell behind the neck, with the legs bent so that the heels are about 6 inches from the buttocks. Hook the feet under some stable object or have them held down by a partner. *Movement:* Curl the head and shoulders and sit up as far as you can go. Return in an uncurling manner so that the lower back, upper back, shoulders, and dumbbell touch the floor in that order. For this exercise, exhale as you come up, and inhale as you go down.

11. *Trunk Rotation: Starting Position:* Stand erect, feet comfortably apart, bar in the shoulder rest position, using a regular overhand grip, with the hands widely spaced. *Movement:* Alternately rotate the shoulders and trunk to the extreme right and then to the left.

12. *Side Bends (with dumbbell): Starting Position:* Stand erect, feet together, with a dumbbell in either hand, arms hanging. Bend as far to the side of the dumbbell as possible. *Movement:* Bend to the opposite side as far as possible, keeping the head and back straight, and return to the starting position. Exhale as you move from the weighted side and inhale as you go the opposite way. Use the free arm as a guide by sliding

it up and down the leg. When completed, switch the dumbbell to the other hand and repeat on that side.

13. *Half-Squats: Starting Position:* Stand erect, feet comfortably apart for balance with the bar at the shoulder rest position, regular grip, wide spacing. *Movement:* Keeping the feet flat on the floor, bend the legs and lower the hips until the knee joint just approaches a right angle, then return to the starting position. Though there will be a slight incline forward as you go down, keep the head and back in a straight line and avoid coming up on the toes. If you have difficulty, place a barbell plate or piece of wood under the heels. Inhale as you lower, exhale as you come up.

14. *Toe Raises: Starting Position:* Same as for Half-Squat. *Movement:* Raise up on the toes as far as possible, and return. Exhale as you come up, inhale as you come down.

15. *Bent-Leg Dead Lift: Starting Position:* Feet well up under the bar, hips and legs bent so that the trunk is at about a 45 to 60 degree angle, hands grasping with an overhand grip, shoulder width apart, with the arms fully extended. *Movement:* Keeping the head and back straight, straighten the legs and bring the bar to the standing thigh position. When you finish this exercise, complete it with 10 *Shoulder Shrugs:* Raise the shoulders as far as they will go, and then bring them forcefully backward, down and back to the starting position. Exhale as you come up, inhale as you go down.

You may wish to use the Weight Training Guide at the end of this chapter (p. 140) to record your progress in weight training.

isometric exercises

Isometrics first gained popularity when Charles Atlas published his commercial program of "Dynamic Tension" in the 1920s. Since then, a limited amount of research has pointed the way to the development of isometrics and their potential in an exercise plan. Essentially, isometric contractions done on a regular basis will build muscle strength. There is little evidence to support claims that muscle endurance is also a prime benefit, though some gain can be expected as strength increases.

advantages and disadvantages

Isometric exercise is a valuable supplement to physical fitness programs because of its simplicity and quick results. In addition, it requires little time and no special equipment is needed. A full program of muscle contractions can be accomplished in five to eight minutes at almost any time or place and in any type of clothing. Another appeal is that it causes little fatigue and no sweating.

Disadvantages are that little or no contribution is made to muscular and cardiovascular endurance, range of movement is not developed, the possibility of joint damage is increased, and a sudden rise in blood pressure during forceful contractions can be potentially dangerous. Because of this latter disadvantage, many doctors and physiologists caution middle-aged and older persons against using isometrics.

principles and practices

Based on the research done on isometric exercises, the following principles and practices are recommended:

1. Use a balanced program that works all the major muscle groups.
2. Hold each contraction for six to eight seconds.
3. Use a maximum effort for the contraction.
4. Do at least one set per day, every day.
5. Use the isometric exercise program in conjunction with another recommended physical fitness activity.
6. Vary the exercises each week and where the body, arms, or legs are flexed, change the angle of flexion so that strength is developed in at least three points through the entire range of motion.
7. In doing isometric contractions, it is important to inhale deeply before each contraction, hold the contraction for the recommended time span, and then exhale.

a basic isometric program

A. *Chair routine*

NECK

1. Clasp hands on front of the head with elbows out to the side. Push forward with the head and resist with the hands.
2. Clasp hands on back of head with elbows out to the side. Push backwards with the head and resist with the hands.
3. Place one hand on the side of the head. Push out with the head and resist with the hand. Repeat on the other side.

ARMS AND SHOULDERS

1. Grasp the chair at the sides and pull up, straightening the head, trunk, and arms while driving the buttocks and back of the legs into the chair.

2. Place the hands on the chair seat at the sides, extend the arms and push down while keeping the head and trunk straight.

3. Place the fist of one hand into the palm of the other at shoulder level, with the elbows extended to the side. Press with the fist while resisting with the hand. Reverse hands and repeat.

UPPER LEGS (SIDES)

1. Place the knees about one foot apart and place the hands on the sides of the knees with the arms extended. Press out with the knees while pressing in with the hands.

2. Turn the palms out and place the hands on the insides of the knees with the arms extended. Press in with the knees and out with the hands.

B. *Floor exercises*

ABDOMINALS

1. Lie on the back with the legs bent, feet flat on the floor, about 6 to 12 inches from the buttocks, and hands on the hips. Raise the trunk halfway to a sitting position and hold. As strength increases, added resistance can be obtained by holding an object behind the head.

BACK (SKIP IF LOW BACK PROBLEM)

1. Lie face down with legs straight and arms at the sides. Raise the legs and trunk off the floor about halfway and hold. As strength increases, raise higher. Variations could include raising the legs only or trunk only.

ARMS AND SHOULDERS

1. Lie face down with the hands under the shoulder joints and legs straight. Keeping the body straight, push to a halfway push-up position and hold. Variations would include holding at different positions, such as three-quarter or one-quarter of the way up.

C. *Standing exercises*

LEGS

1. Stand with the heels about 12–15 inches away from a wall, with the buttocks, back and head pressed against it. Lower the body, letting the heels raise until the thighs are parallel to the floor and then hold. Variations include holding on one leg alternately and changing the held position to various levels less than parallel.

BACK, ABDOMINAL, AND GLUTEAL

1. Stand with your back against a wall, heels about 4 inches out from the base and your knees slightly bent. Simultaneously tighten the abdominals and squeeze the gluteals so the pelvis moves forward, press as much of your back as possible against the wall, and hold.

ALL-MUSCLE TENSOR

1. To finish the workout, raise up on the toes, pull the stomach in, squeeze the buttocks hard, raise the shoulders, extend the arms outward to the side, and force full contraction of all muscles for ten-seconds.

iso-kinetic exercises

There have been several efforts to combine the advantages of isotonic exercises, such as calisthenics and weight training, and isometric exercises for the purpose of gaining maximum strength in a short period of time while still developing muscle endurance and maintaining a range of motion for flexibility. One of the most successful of these efforts is the Exer-Genie exerciser, the commercial exercise device described in Chapter 4. It is exceptionally adaptable to individual exercise and is also widely used in many high schools, colleges, and universities throughout the country, as well as by professional athletic teams.

the Exer-Genie exerciser—advantages and disadvantages

Advantages of the Exer-Genie exerciser are: (1) Muscle groups can be individually isolated and worked by a variety of special exercises. (2) Exercises can be adapted from highly sophisticated weight training equipment or barbell training equipment and simulated with the Exer-Genie exerciser. (3) Specific body conditioning for various sport techniques can be combined with technique training, since the various movements can be simulated and performed under controlled resistance. (4) The equipment is relatively less expensive than weight training equipment, much more easily stored and carried, and can be hooked up and used anywhere. (5) The time needed for a complete exercise routine comparable to weight training is considerably less. (6) It combines the advantages of maximum strength development from isometric contractions with the range of motion that weight training provides, but with the added benefit of maintaining a steady resistance throughout the entire range of movement.

Like weight training, however, a major disadvantage is the lack of sufficient cardiovascular endurance development. But it is already apparent that a full program is made up of many parts, and the Exer-Genie exerciser can be one of these parts. Other disadvantages, absent in weight training, are some difficulty in controlling and measuring resistance, the inability to keep a precise record of progress, and a little greater difficulty in developing the techniques. Psychologically, another disadvantage to some people is not feeling you've had a long enough workout.

For a full description of the Exer-Genie exerciser and its uses, refer to the Exer-Genie Exerciser Instruction Manual in the Bibliography. Presented here are two of the basic programs used with the device.

six-minute program for general conditioning

This program combines a set of exercises done in sequential order, called the Big Four (see Fig. 7), plus two additional exercises called the Bicycle and Row which are to be done daily. It is recommended that the Big Four be repeated three times, pausing a minute or two between each repetition or until breathing becomes normal and the pulse rate drops to below 120 (a ten-second count of 20).

Fig. 7.   Exer-Genie Big 4

The Big Four is begun with a ten-second isometric hold obtained by applying finger pressure on the nylon cord looped over the handle. Prior to the exercise, a predetermined resistance is set against the cord pull through the device by means of a revolving dial. This pre-set resistance is constant as long as no other pressure is placed against the rope movement. But when strong pressure is applied against it, as in pressing it against the handle with the finger, the rope becomes impossible to move, regardless of the pre-set resistance. Following the isometric hold, finger pressure is released gradually and the following exercises are performed in sequence: Dead Life, Leg Press, Biceps Curl or Upright Row (depending on hand position; i.e., Upright Row is bringing the hands straight up to the chin), and the Standing Press, finished by raising up on the toes. Following the Leg Press, the cord is dropped to the floor and the amount of resistance is governed by the amount pre-set and the length of rope on the floor. By the time the Standing Press movement is to be performed, the resistance has increased again from the smaller amount needed for the Biceps Curl or Upright Row to the larger amount required for the Standing Press, as a result of the added cord length out and the floor resistance against its movement.

Following the Big Four, one more minute of additional exercise completes the Six-Minute Program; this is either the Bicycling or Rowing Exercise, alternated daily. The Bicycling Exercise is for the legs and buttocks and the Rowing Exercise is for the shoulders, upper back, and abdominals.

program for exercising special muscle groups

A general program for exercising special muscle groups and for total body development can be developed from the list of exercises shown in Table 5 and explained fully in the Exer-Genie Exerciser Instruction Manual. The program would take a little longer, but would be more complete and a better daily or even every-other-day program than the Six-Minute Program. All the exercises are performed by first holding a ten-second isometric contraction (obtained by finger pressure as described in the Big Four) followed by the gradual release of pressure and complete movement, with resistance governed by the pre-set dial amount and cord manipulation.

aerobics

Some discussion has already been held concerning the nature of aerobic activities. The man responsible for popularizing the term *aerobics* is former Air Force Major Kenneth Cooper, M.D., author of the widely

table 5
program for exercising special
muscle groups*

Arms:
1. Triceps Pull
2. Biceps Curl

Upper Body:
1. Lats Pull (overhead pull down behind the neck)
2. Dry Land Swim
3. Rowing Exercises (listed under exercises for the abdomen, midsection, and lower back)
4. Standing Bench Press

Neck:
1. Neck Pull Forward
2. Neck Pull Backward

Chest:
1. Standing Arm Pull (from side extended arm position across body)

Side:
1. Side Bend

Abdomen, Midsection, and Lower Back:
1. Sit-up
2. Forward Bend
3. Rowing
   a. Upright Row
   b. One-Man Row (sitting rowing motion)
   c. Two-Man Row (with partner)

Legs and Hips:
1. Bicycle
   a. Straight Leg
   b. Bent Leg
2. Thigh Pulls
   a. Outer Thigh Pull (arm extended leaning position against wall and swinging leg out)
   b. Inner Thigh Pull (same, except swing leg across and in front of the other one)

---

* Adapted by permission of Exer-Genie, Inc., P.O. Box 3237, Fullerton, California 92634.

read books, *Aerobics* and *The New Aerobics*. Aerobics is simply a term used to describe a group of exercise activities that have a particular beneficial effect on the condition of the heart, lungs, and blood vessels as well as on overall health and fitness. It is a group of activities that conditions the heart, lungs, and vascular system in a way unattainable in other types of activity. The activities generally accepted as aerobic are running, jogging, walking, cycling, swimming, and hiking. Also included are some sports and games such as handball, squash, tennis, basketball, soccer, and similar activities when they are pursued in a manner that has a conditioning effect on the heart, lungs, and blood vessels.

Jogging, running, and walking have been given the strongest emphasis in popularizing aerobic programs. Although men have been running and jogging since the time of the early Greeks, recent recognition came because of new attention to physical fitness programs throughout the world.

In the United States, William Bowerman, track coach at the University of Oregon, directed his attention to jogging and running as a result of witnessing its effect on New Zelanders during a visit to New Zealand where jogging is almost a way of life. When Bowerman returned to the United States and praised jogging as a physical fitness activity suitable for all ages, people began to listen. His reputation as a fine track coach helped him become a leader in the jogging movement through his own book, *Jogging*. (See Bibliography.) Others followed and research began to substantiate the claims that jogging is a healthy way to physical fitness. Dr. Lawrence Lamb of Baylor University who was also associated with the NASA program, along with Cooper, supported through medical research the feasibility of aerobic activities as programs for optimum fitness.

There are many sound programs available for consideration in selecting a personal approach to an aerobics program. The YMCA "Run for Your Life" program is an excellent example of one. So is Bowerman's jogging program. Many school districts and universities hold adult fitness courses that emphasize aerobic activities, and throughout the country "jogger's clubs" have been formed that are open to both men and women. Dr. Harry J. Johnson, M.D., has recently written a book called *Creative Walking* (listed in Bibliography) that maps out a program for those who would walk their way to fitness. It is popular with older persons and is worth consideration by the young who could profit from its philosophic approach. You can adopt any of these programs or you can devise your own by applying the principles and using the techniques explained under General Endurance Training in Chapter 4. Though all the aerobics programs mentioned are sound, one of the simplest and most inclusive is Dr. Cooper's program.

cooper's aerobics program

What makes Cooper's program both scientifically accurate and simple is that, through extensive laboratory tests, Cooper measured the oxygen consumption cost of various aerobic activities performed at various intensities (duration, rates of speed) and converted these into an equivalent "point" system. For example, running or walking one mile in between 14:30 and 20:00 minutes costs approximately 7 milliliters of oxygen per kilogram of body weight per minute (ml/kg/min) and is given the equivalent of 1 point, while running it in between 6:30 and 8:00 minutes costs approximately 35 ml/kg/min and has a value of 5 points. Other tests, including the 12-minute run-walk test, established how many points a week were needed to produce and sustain a good minimum training effect; this was set as the maintenance goal to be reached. The goal arrived at is 30 points a week, earned through any of the activities given, over a minimum of four days and up to six days (seven days optional). Thus, one could choose to do either a 7 1/2 point activity for four days or a 5 point activity for six days in order to earn the 30 points for the week.

Based on Fitness Categories for various age groups as established by the 12-minute test, detailed and gradually progressive conditioning programs leading to the 30 point goal are outlined for walking, running, stationary running, swimming, cycling, handball, squash, and basketball activities. For those who fall below Category IV on the 12-minute test, the program leading to the maintenance goal is divided into a Starter (or preconditioning) program and a Conditioning program over a total period of from ten to sixteen weeks, depending on the Test Category and activity selected. An example of the Running Exercise Program for those under thirty years of age and in Fitness Category I (very poor) is shown in Table 6. For a complete explanation of the many other programs outlined, refer to the books listed in the Bibliography.

advantages and disadvantages

The advantages of Cooper's program include its combination of scientific measurements and simplicity—though some top physiologists question the accuracy of his statistics—clearly established goals, gradual level of progression, and variety. Like any other aerobics plan, the major advantage is the development of cardio-respiratory endurance and the beneficial effect on the heart, lungs, and entire vascular system. Since aerobic activities cost the most energy, they offer an excellent way to control body weight, get rid of unsightly and dangerous fat, and streamline the body, as well as to rid oneself of nervous tension. Muscle tone and general muscle endurance are also improved, along with a certain

table 6

running exercise program
*(under 30 years of age)**

		Starter**		
Week	*Distance (miles)*	*Time (min.)*	*Freq/Wk*	*Points/Wk*
1	1.0	13:30	5	10
2	1.0	13:00	5	10
3	1.0	12:45	5	10
4	1.0	11:45	5	15
5	1.0	11:00	5	15
6	1.0	10:30	5	15

** Start the program by walking, then walk and run, or run, as necessary to meet the changing time goals.

conditioning
fitness category I
(less than 1.0 mile on 12-minute test)

Week	*Distance (miles)*	*Time (min.)*	*Freq/Wk*	*Points/Wk*
7	1.0	9:45	5	20
8	1.0	9:30	5	20
9	1.0	9:15	5	20
10	1.0	9:00	3	21
	and			
	1.5	16:00	2	
11	1.0	8:45	3	21
	and			
	1.5	15:00	2	
12	1.0	8:30	3	24
	and			
	1.5	14:00	2	
13	1.0	8:15	3	24
	and			
	1.5	13:30	2	
14	1.0	7:55	3	27
	and			
	1.5	13:00	2	
15	1.0	7:45	2	30
	and			
	1.5	12:30	2	
	and			
	2.0	18:00	1	
16	1.5	11:55	2	31
	and			
	2.0	17:00	2	

* From *The New Aerobics* by Kenneth H. Cooper, M.D. Copyright 1970 by Kenneth H. Cooper, p. 55. Reprinted by permission of the publisher, M. Evans and Company, New York, New York.

amount of general strength resulting from improved circulation and healthier muscle tissue.

Specific disadvantages of the Cooper program are few, but warrant attention. Because it is a program designed for mass public use and older groups, of necessity, it must be ultraconservative in its starting points and rates of progression. Though caution should be used, these starting points and progression rates can be adjusted and accelerated by using the Heart Rate Monitoring system described in the last chapter. That is, after beginning at the recommended starting point for your age and Fitness Category, the heart rate check can determine whether or not you can raise the starting point and how long you should remain there before progressing to the next level. The check may very well indicate that you should spend a longer time at a particular level than that shown on the training chart. It can serve, then, as a valuable safety device as well as a method of providing greater flexibility. For older persons, experience has shown that at about the time when the mile run is cut to between eight and ten minutes, it is better to progress by lengthening the distance and keeping the rate constant rather than by attempting to cut the running time further.

A final precaution, as Cooper himself mentions, is that because of the specific training effects principle, it is best to remain with one type of activity until the maintenance goal is reached. After that, the maintenance program may include a variety of activities as long as the 30 point minimum a week is earned.

Important as aerobic activities are, they are still not the complete answer to physical fitness through exercise. Other elements of motor fitness, such as flexibility and coordination are needed at all ages. Qualities such as specific local strength, muscle endurance, and power must be maintained at a minimal level and are required at higher levels for those engaging in certain types of work or sports activities. These are best developed through one of the other anaerobic programs discussed.

You can record your progress in the aerobics program on the Aerobics Running Program Guide at the end of the chapter, p. 141.

the royal canadian air force exercise plans

For those who want a good, well-balanced, and challenging workout in a minimal amount of time and with the most convenience, the RCAF 5BX and XBX exercise plans (listed in Bibliography) have proven to be excellent programs. Both were developed in the late 1950s, and were adopted as the official exercise plans for RCAF desk personnel following extensive testing and revisions. Their popular appeal throughout Canada and the United States is that they require little time and space, no equipment, and can be done anywhere. The plan tells its followers what to do, where to start, how fast to progress, and how to maintain the particular level of conditioning reached. The XBX is a ten-exercise program for women based on a twelve-minute workout and the 5BX is a five-exercise program designed for men and structured as an eleven-minute workout.

the 5BX plan

The 5BX plan is composed of six charts containing five basic exercises. They are always performed in the same order, guided by specific time limits, but must be completed within the total time period allotted (eleven minutes). The exercises include: a flexibility exercise (two-minute limit), a sit-up exercise (one minute), a back arch exercise (one minute), a push-up exercise (one minute), and running-in-place for a certain number of counts (six minute limit). A mile-run or two-mile walk within separate time limits may be substituted for the stationary run.

Each chart has twelve levels indicating the number of exercise repetitions to be performed within the specific time limits (time goals if walking or running is substituted). At each higher level, the number of exercise repetitions is increased (speed overload). The starting level on Chart I is determined by age, and progress to the next level is made only when all exercise repetitions are completed within the total eleven minutes allowed. The procedure is continued through the charts until a predetermined Chart and level maintenance goal is reached—again based on age. At this point, workouts are performed three days a week instead of daily. As progress is made from one chart to the next, a slight variation in the basic exercises is made in order to make them more difficult to execute (resistance overload). Though the exercises and repetitions on Chart I are easy enough for the youngest child or oldest person, completing those on Charts V and VI within the eleven-minute time limit is challenging enough for the best conditioned athlete.

the XBX plan

The XBX plan for women follows the same basic procedure as the 5BX, except that the plan is composed of four charts of ten exercises arranged in progressive order of difficulty. Each chart again is divided into twelve levels. Though the stationary run segment is similar to the 5BX, the other exercises are aimed more at the development of the waist, hips, legs, and ankles.

advantages and disadvantages

Each program is sound and practical with the age progression approach. Advantages are: (1) It requires no other person to assist. (2) No special equipment is needed. (3) Only a small amount of time is needed. (4) It provides a maintenance program when the desired fitness level is reached. (5) It allows for retrogression to previous levels if workouts have been missed. Disadvantages are few, but include: (1) Overall fitness goals are limited to what can be achieved in eleven or twelve minutes. (2) It does not provide for strong cardiovascular conditioning in the stationary run segment. (3) It is ultraconservative in establishing the starting level. (4) The various motor fitness elements and cardiorespiratory endurance are equated on the same level.

Obviously, arm and shoulder strength cannot be equated with flexibility in terms of difficulty and rate of development. Neither is there a high correlation between the other components. Basing overall progression from level to level on the basis of minimum performance in each specific exercise, therefore, is not valid. Being low in one area should not delay progression or take away excellence in another. This weakness can easily be remedied by following the plan exactly as it is, but progressing to the next level on an individual element basis and recording progress by circling or checking each level reached. Another criticism of the program is the advisability of simultaneous trunk-leg raise exercises for those with low back problems or "swayback." Substituting another back strengthener or flexibility exercise is recommended for these persons. (See Back Stretch P-F Plan, Exercise 3, p. 135.)

the P-F plan

The P-F (physical fitness) plan is an adaptation of the 5BX program prepared by the Health and Physical Education Department at Duke University and modified slightly on the basis of our regular and adult class experiences. The plan is based on an eleven-minute daily workout which includes five basic exercises and follows the same theory as that of the RCAF programs.

The P-F levels are based on of percentile ratings. The recommended starting point for young persons in fairly good condition is Level 10. For those in fair condition, Level 6 is recommended, while those in poor condition should start at the Level 2. Older persons are urged to start at the first level or as recommended by their physician, and then raise it if they find it to be too easy. For those who find even this level too difficult, a simplified exercise program (P-F Preparatory) is recommended until it can be done without undue fatigue. This program includes five simple exercises and walking. Both programs are explained below and the progression chart for the P-F plan is shown in Table 7. The advantages and disadvantages are the same as for the 5BX and XBX plans and can be modified accordingly.

simplified P-F preparatory program

1. *Shin Touch:* Bend forward and touch shins—15 times.
2. *Half-Squat:* Bend knees and go to a half-squat position and return. Repeat 15 times.
3. *Wall Push-Ups:* Stand 2 feet away from the wall. Lean on hands extended against the wall, touch nose to the wall, and return. Repeat 25 times.
4. *Head Lift (front):* Lie on back; lift head 6 inches. Repeat 10 times.
5. *Head Lift (back):* Lie on stomach; lift chin 6 inches. Repeat 10 times.
6. *Walking:* Walk at least three blocks. (Though not stated in the plan, a stationary run may be substituted, beginning with about 2½ minutes duration at a 70–80 step/minute rate and extending it to 5 minutes by the third week).

P-F plan exercises

1. *Toe Touch:* Feet astride, arms upward. Touch floor 6 inches outside left foot, between feet, and then 6 inches outside the right foot; come up, bend backward as far as possible, and repeat. Reverse direction after half the repetitions are completed. Keep legs straight.
2. *Sit-Ups:* Back lying position, legs bent, feet together, hands clasped behind the head. Sit up and alternately touch right and left elbows to knees.
3. *Back Stretch:* Back lying position, legs straight. Tuck the head, bring both knees to the chest, clasp hands around the knees and pull tightly. Return. (No back problem: See Trunk Raise, p. 111.)
4. *Push-Ups:* Standard. (See Calisthenics, Exercise 4, p. 110).
5. *Stationary Run:* Count a step each time the left foot touches the floor. Lift the feet approximately 4 inches. After every 75 counts, do 10 Knee Bends. Repeat this sequence until the required number of steps is completed. Perform the knee bends as described previously; i.e. to just short of a 90 degree angle.

water exercises

Swimming as an aerobic activity has already been discussed. Because of its popularity, the increased availability of both public and private pools, and its suitability for even the handicapped, some mention should

table 7

p-f plan: activity level chart**

Exercise

* Level	Toe Touch	Sit-Ups	Back Stretch	Push-Ups	Stationary Run
	1	**2**	**3**	**4**	**5**
24	52	50	50	46	600
23	51	49	49	44	590
22	50	48	48	42	580
21	49	47	47	40	570
20	48	46	46	38	560
19	47	45	45	36	550
18	46	44	44	34	540
17	45	43	43	32	530
16	44	42	42	30	510
15	43	41	41	28	490
14	42	40	40	26	470
13	41	39	39	24	450
12	39	37	37	22	430
11	37	35	35	20	410
10	35	33	33	18	390
9	33	31	31	16	380
8	31	29	29	14	360
7	29	27	27	12	340
6	27	25	25	11	320
5	24	22	22	9	300
4	21	19	19	8	280
3	18	16	16	7	260
2	15	13	13	6	240
1	12	10	10	5	220
Minutes for → Each Exercise	2	1	1	1	6

Rating column (left margin): *Good*, *Average*, *Below Average* — *Faculty*, *Students*
Level groupings: *Good*, *Average*, *Below Average*

* Based upon norms for university male freshmen. Adapted norms for faculty are indicated.

** From *P-F: A Physical Fitness Plan*, Department of Health and Physical Education, Duke University, West Campus, Durham, North Carolina, 1969.

also be made of a complete water exercise program that can be used not only for circulatory endurance, but also for muscular strength, endurance, and flexibility.

The program is one devised by C. Carson (Casey) Conrad, Chief of California's Bureau of Health Education, Physical Education, Athletics, and Recreation and Executive Director of the President's Council on Physical Fitness and Sports. The program consists of a Low Gear (ten minutes), Middle Gear (thirty minutes), and High Gear (fifty minutes) series of exercises that includes various Flutter Kicks, Knee Raises, pedaling, and extension movements combined with bobbing, treading, and lap swimming intervals. It is an excellent all-around program and worth considering if you prefer your exercise in a more refreshing environment. Details of the program can be obtained from the reference listed in the Bibliography.

guides for heart rate monitored exercise and measuring improvement

a. exercise heart rate

A safe maximum Exercise Heart Rate for a sedentary person just beginning a training program is one that does not exceed 70 percent of the Working Heart Rate (difference between Maximum Heart Rate and true Resting Heart Rate). For an effective cardiovascular training effect, the HR should be elevated to at least 60 percent of its WHR and maintained there for three minutes or longer. The higher and longer it is maintained, the greater the training effect. For young persons, and older persons who have been exercising regularly for at least six weeks, an

	Minimum Exercise HR	*Maximum Exercise HR*
Dates:	__ __ __ __ __	__ __ __ __ __
MHR (Men: 220–age) (Women: 226–age):	__ __ __ __ __	__ __ __ __ __
Resting HR, A.M., lying: (subtract)	__ __ __ __ __	__ __ __ __ __
Working HR:	__ __ __ __ __	__ __ __ __ __
Percentage: (multiply)	__ __ __ __ __	__ __ __ __ __
Resting HR (add)	__ __ __ __ __	__ __ __ __ __
Exercise HR–1 min.	__ __ __ __ __	__ __ __ __ __
Exercise HR–10 sec.	__ __ __ __ __	__ __ __ __ __

Exercise Heart Rate range of 70 to 80 percent may be used. Sustaining Maximum Heart Rate should always be avoided. For monitoring the HR during an aerobic or any other program, use a ten-second pulse count taken within five seconds or less after an exercise bout. Remember, as you continue your program, it will take a greater exercise intensity to reach your minimum and maximum levels—a sign of improvement.

b. recovery heart rate

Two minutes after any exercise bout, your heart rate for a ten-second count should drop to 20 or below (21 for women). A higher count would indicate the bout was too strenuous and call for an adjustment. Knowing your "normal" ten-second Resting Heart Rate, sitting and standing, and noting how long it takes for full recovery is another good guide and measure of improvement. They can also be used as checks throughout the day.

Dates: __ __ __ __ __   __ __ __ __ __
RHR, sitting   __ __ __ __ __   RHR, standing   __ __ __ __ __
   (10-sec.)   (10-sec.)

warm up and calisthenic exercise guide

warm-up exercises

1. Jumping Jack (15)
2. Side Bender (5 ea. side)
3. Side Twister (5 ea. side)
4. Forward Bend & Stretch (5)
5. Elbow Thrust (15)
6. Shoulder Roll (5 ea. direction)
7. Neck Roll (3 ea. direction)
8. Knee Lifts (10 ea. leg)
9. Knee Bends (10)
10. Toe Touch (10)
11. Toe Raises (10)
12. Ankle Roll (10 ea. way)
13. Slow Jog in Place (50 counts)
14. Deep Breathing (5)

Decrease reps in warm weather, increase in cold

top to bottom exercise record

Exercise	Rec. Reps	Date and Performance Record
1. Toe Touch Series	3–6 ea.	
2. Knee Bends	10–20	
3. Sprinter's Drive	5–20	
4. Push-Ups	5–20	
5. Sitting Stretch Static	5–10	
6. Sitting Stretch Ballistic	5–10	
7. Gather Sit-Ups	10–30	
8. Leg Cross-Over (sitting)	5–20	
9. Side Leg Lifts	5–20	
10. Back Leg Lifts	5–20	
11. Trunk Raises	5–20*	
12. Flutter Kick	10–40*	
13. Jog & Run In Place	6–12 inter.	

* Skip if low back problem and see Remedial Exercises for Low Back (p. 161)

Notes: (Ease or difficulty; Exercise HR, Recovery HR, etc.)

weight training guide

Exercise	Starting Load	Rec. Reps	Date and Performance Record
Chest			
Bench Press	1/2	5–10	
St. Arm Pull-Over	1/6	5–10	
Lat. Raises, Lying	10 lb. ea.	10–20	
Arms-Shoulders			
Stand. Press	1/4	5–10	
Biceps Curl	1/5	5–10	
Triceps Curl	1/5	5–10	
Back			
Bent-Over Row	1/4	5–10	
Bent-Over Lat. Raises	10 lb. ea.	10–20	
Forward Bend	1/5	5–10	
Waist			
Bent-Knee Sit-Ups		10–20	
Trunk Rotation	1/5	10–20 ea. sd.	
Side Bends		10–20 ea. sd.	
Legs			
Half-Squats	1/2	10–20	
Toe Raises	1/2	10–20	
Bent-Knee Dead Lift	1/2	10–20	

Starting Loads = % Body Weight.
Recommended Reps are for Double Progression Method of Training: 1 to 3 sets.
For Hypertrophy: 3 sets at 1/2 10RM, 3/4 10RM, and Full 10RM loads.

Record Performance as follows:

$\dfrac{\text{Sets–Reps}}{\text{Load}}$ Example: $\dfrac{1\text{–}5}{90}$ would mean you performed one 5-rep set with a 90 lb. load

Notes:

aerobics running program guide

Maximum HR: _____ Exercise HR Range: _____ (10-second guides)

Date	Distance & Method	Time	EHR*	2-min. RHR	5-min. RHR

* Record 10-second EHR immediately after finishing run to serve as a guide for adjusting subsequent runs. If jog/walk routine, make 10-second spot checks and record last one.

Maintenance Goal:　1 mile in under 8 minutes; 6 times a week, or
1 mile under 6:30 5 days a week, or
1½ miles in 12:00–14:59, 5 times a week

6

posture
and low back
problems

Good static and dynamic posture depends on good body mechanics in assuming various positions or performing certain actions. By good body mechanics, we mean the proper alignment of body segments and a balance of forces so as to provide maximum support, the least amount of strain, and the greatest mechanical efficiency.

In this chapter we will discuss some of the basic principles and rules relating to good body mechanics in assuming the various postures associated with everyday activities. Included will be a description of the essential elements of good body mechanics in the standing, walking, sitting, and lying positions and in performing such common tasks as stooping, lifting, carrying, pushing, and pulling. It is almost tragic how many persons suffer needless aches, pains, and even temporary disability from either ignorance of proper body mechanics or just plain carelessness in performing these relatively simple acts. With respect to postural deviations and low back ailments, we will also discuss their principal causes, possibilities for correction, and specific remedial exercises for them.

The essence of good body mechanics is a balance and application of forces in a position that will provide maximum support, the least amount of strain, and the greatest mechanical efficiency, i.e., minimum muscular effort. It is assuming a position that enables the body to act most effectively with the least amount of interference with internal functions and the least strain, tension, and possibility of injury. In addition, gracefulness rather than rigidity should characterize all good body mechanics.

The first step in developing good posture habits is having a clear mental image of what the standards are. Once established clearly in the mind, correct posture and body mechanics must then be stressed in all phases of daily living if it is to become habitual. It cannot be attained by practicing only a few minutes a day. You must become "posture conscious" and back it with a firm desire to recognize and correct faults until good posture (static and dynamic) becomes as automatic to you as breathing.

standing posture

Your standing posture is basic to every other position. It reflects the way you habitually carry yourself not only when standing, but when sitting or moving about. Improper alignment in the standing position inevitably carries over to other postures; it is therefore here that all corrective efforts must start. If through conscious effort, proper alignment still can not be attained, then a deviation is definitely indicated and other corrective measures must be taken.

At the start, you should recognize that there is no "picture perfect" posture for all persons. Differences in body build and other inherited characteristics of growth and development preclude any precise patterns. There are, however, certain generalizations that can be used as standards for all.

Proper basic alignment. In general, the skeletal frame is aligned so that the weight of the body can be supported with a minimum of dependence on the muscles. For proper balance, distribute the weight evenly around a vertical gravity line passing directly through the midpoint of the body, with the center of gravity (approximately the middle of the pelvic area) directly over the supporting base. (See Fig. 8.) From a side view, correct alignment would enable an imaginary line to pass through the lobe of the ear, the middle of the shoulder and hip, slightly back of the knee cap, and to the front of the outer ankle bone. (See Fig. 9.)

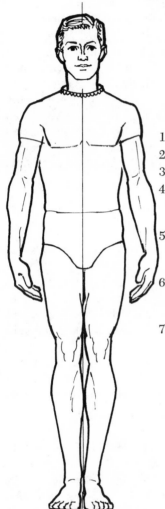

1. Body tall as possible without strain
2. Head and neck centered between shoulders
3. Shoulders relaxed; fall evenly
4. Pelvis level (top of hip bones even) and centered squarely above feet, providing firm support for the trunk
5. Weight evenly distributed on both feet and centered between the heels and balls of both feet
6. Feet pointed forward or slightly outward, with ankles neither sagging inward nor excessively cupped outward
7. Center of gravity formed in a vertical line through the midpoint of the body situated in the middle of the pelvic area directly over the supporting base (feet)

FIG. 8.   Standing Posture (Front View)

1. Body stretched upward tall as possible without strain

2. Head erect, eyes level with chin and neck straight

3. Chest moderately elevated without strain and furthest point forward

4. Abdomen flat, but not drawn in to extent that normal breathing is restricted

5. Shoulders held backward and downward, but not tense

6. Spinal curves gentle and not exaggerated at the base of the neck, upper and lower back

7. Buttocks slightly contracted and drawn down and under, so that the pelvis is tilted slightly upward in front

8. Knees straight and relaxed with no evidence of stiffness

9. Weight evenly distributed and balanced just back of the middle of the feet, with the center of gravity in the middle of the pelvic area

10. Body segments aligned so that the ear lobe, tip of the shoulder, middle of the hips, back of the knee cap, and front of the ankle bone are in a vertical line

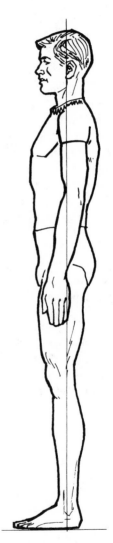

FIG. 9. Standing Posture (Side View)

When this alignment is disturbed by faulty posture positions of one or more joints, the entire body is thrown out of line by a compensatory deviation of other parts. For example, if the head drops forward, there is a tendency for the upper back to curve more than normal and, in turn, this creates an increased curvature of the lower back, tilting the pelvis down in front. The muscles must then overwork to counteract the pull of gravity producing fatigue and unnecessary strain. In addition, a general slovenly appearance results.

Developing good basic alignment. The simplest way for getting the body into a good standing position is to stand with the weight solidly and evenly distributed on both feet and stand tall—that is, "stretch" the body upward as tall as possible, but without strain. In doing this, however, don't tilt the head or raise the shoulders. By keeping the eyes level and pressing the neck back so that the curve of the neck is flattened, this tendency can be avoided. Draw the chin inward so that the point is carried directly over the notch at the top of the breastbone.

In this position, provided there are not structural or muscular defects, the rest of the body segments will fall almost naturally in line. The various antigravity muscles of the body which keep the body in an erect position, will all contract in their natural holding position. The chest will raise automatically; the muscles of the back, buttocks, and legs will contract, and the stomach will flatten and contract normally.

At first, "standing tall" will feel unnatural and require constant, conscious attention. But, in time, it will become something you do without thought. The physical and psychological rewards you will receive will make the effort well worthwhile. Another excellent way to develop the major postural muscles is to do Isometric Exercise C-1 (under *Back, Abdominal, and Gluteal*), on page 124, four or five times a day.

walking

Easy, efficient, and graceful walking is merely an extension of good standing posture, with the added movement of the legs for propulsion and the swinging of the arms to maintain a counteracting balance. Keep the head, neck, shoulders, trunk, and hips in the same alignment and balance as in the standing position, free from tenseness. Remember to "tuck in" the buttocks and to "stand tall," but there is no need to accentuate any other position, such as keeping your chin in, thrusting the shoulders back, the chest out, or pulling the stomach further in. If you do, rather than an easy, graceful walk, it will look more like that of a wooden soldier and actually cause more fatigue.

The legs, rather than being considered as supports for the body, should now be thought of as pendulums which swing from their sockets in the pelvic girdle in a free, easy motion. As you move forward, keep the feet pointed straight ahead or slightly outward and let them pass close together. The length of the step should be comfortable and allow the knees to be slightly flexed as the legs swing forward and the feet strike the ground. Let the midpoint of the heel strike the ground first and then the ball of the foot. By striking the ground in this manner, swinging from the hips, and keeping the knees slightly flexed, you'll walk with an easy grace and a light spring that sends the weight forward.

sitting

Correct sitting posture involves the same alignment and balance from the head to the hips as in the standing position. It is important to sit well back into the chair and allow the back to follow the natural contour of the chair. This allows for maximum support of the spinal column and for the relaxation of the supporting muscles. It is particularly important that there be no space between the back of the hips and the back of the chair. Avoid the common tendency of allowing the hips to slide forward, since this creates an excessive curvature of the lower spine and places it under an undue amount of strain. Keep the upper part of the legs supported by the chair up to the curvature of the knees, and the feet resting flat on the floor. Naturally, it is important that furniture construction and dimensions be selected so that they allow for correct sitting.

Variations in the sitting position, of course, are both natural and necessary. It would be tiresome and rigid to maintain the same fixed position for a long time. But the variations must be made without disturbing the basic alignment and support of the body weight. Crossing the legs at the ankles, extending the legs forward, and crossing one leg above the knee of the other, are variations that can be made, but without changing the basic position of the hips, trunk, and head. Leaning forward while talking, listening, or watching is also natural, but do this by leaning forward at the hips rather than at the waist, shoulders, or head, so that proper head and trunk alignment can be maintained.

Sitting on the legs should always be avoided, since it curtails circulation. Another tendency to avoid when sitting on the floor or in a soft chair with the legs off the floor is sitting with the legs bent, knees turned inward, and the lower part of the legs turned out. This causes an abnormal dislocation at the hip socket, and in young children especially can cause difficulty in correct leg-hip bone alignment and joint development. When sitting in a leg elevated position, the most natural and best anatomical position is with the legs bent, ankles crossed, and knees falling outward (Indian or cross-legged sitting position).

When working at a desk, follow the same principles of body alignment and foot support as that described for the basic sitting position, but with a forward lean from the hips. Keep the rear part of the hips in contact with the back of the chair and the back straight. Avoid overdropping the head or rolling the upper back. For reading or writing, keep the chair close enough to allow the rear, inner fleshy part of the forearm to rest on the desk without pushing the shoulders up, and the feet flat on the floor. Occasionally pushing back from the desk to a straight position while contracting the stomach muscles or stretching the arms, shoulders, and neck backward can help to relieve built-up muscle tensions.

sleeping position

If tired enough, of course, a person may fall asleep anywhere and in any position. Once asleep, he is likely to change positions a number of times. But the position he assumes may result in awakening not refreshed, with a crink in the neck, pain in the lower back, or an arm completely numb and difficult to move. If this happens frequently, the full benefits of sleeping are not being realized, and it may pay to examine your initial sleeping position habits. A crink in the neck, for example, may be the result of sleeping with the head propped up with a high pillow. A paralyzed arm may be from the habit of using the arm as a support for the head. Pain in the lower back is generally the result of sleeping flat on the stomach, especially on a soft mattress. In any case, the body is being placed in an unnatural position, making it impossible to achieve complete relaxation.

The key to restful sleep is complete body relaxation. The muscles must be made to "let go" as much as possible. This means that the sleeping position must be one that eliminates all unnecessary muscle tension caused by improper body alignment, balance, or restrictive positions. As in the standing or sitting positions, the spinal column should be supported so that excessive curvatures are avoided at the neck and lower back and so that there is the least amount of strain on the supporting muscles and ligaments. Avoid jacknife and other twisted arm, leg, and trunk positions. These not only produce unnecessary muscle strain, but also interfere with the arrangement of internal organs and with circulation. Because the body will move naturally to relieve itself, these positions will only increase the tendency for frequent changes of position which can cause awakening and deprive a person of the restful sleep he needs.

The position which provides the best support, least interference with internal organs, and most freedom from unnecessary tension is the back lying position with a support (pillow) under the knees to slightly flex the knees and hips and keep the lower back pressed firmly against the mattress. In this position, the arms should be relaxed and remain below shoulder level to avoid any undue restriction or tension. The head may be slightly elevated for comfort, but not so high as to produce strain on the back of the neck. Adjustable hospital beds appear to be best for inducing the ideal sleeping position.

For those accustomed to sleeping on their side or stomach, certain precautions should be followed. In the side lying position, draw one or both knees up slightly to relieve strain on the lower back. Keep the underarm away from the body in a bent arm position with the back of the hand resting along the head and the top arm in front of the body for balance. Avoid lying directly on the deltoid muscle (at the top of the arm and just below the shoulder). In tracing complaints of pain in this muscle, we have often found it resulting from this practice. When sleeping on the stomach, a pillow under the stomach will help to keep the back straight. Since you are usually semiconscious during sleep movements, you can train yourself to reach for the pillow and make these adjustments during the night.

stooping and lifting

Care should always be exercised when bending to lift any object, especially if it weighs over ten pounds. Always bend at the knees and let the trunk incline forward at the hips (not the waist), so that the basic alignment from head to hips is kept straight. In lifting a heavy object, the feet should be spread about shoulder width apart and kept flat on the ground for maximum support and balance. Before lifting, grasp the object firmly, with the arms either straight or flexed, depending on the weight of the object. The lift itself should always be made by extending the legs, using the arms and hands merely as attachments, and keeping the load as close as possible to the body and center of gravity. Failing to bend the knees and to lift with the legs or lifting with the arms only places an abnormal strain on the lower back muscles which should simply be used to straighten the trunk. Violations of this lifting technique are a common cause of back pain.

If the object is very heavy, place one foot in front of the other during the lift, so that balance is not lost backwards. In lowering an object to the floor, the action again should be made by bending the knees, keeping the trunk straight and flexed at the hips.

handling an object overhead

In handling an object overhead, such as in placing or removing an object from a high shelf, spread and place the feet in a front and back position for a broader base of support. The hips and body should again be kept in straight alignment. When the object is up past the center of gravity, the weight should be shifted to the rear foot for greater balance. Care should be taken, however, not to lift overhead at too great a speed, since it is more difficult to terminate a backward movement when the object passes the center of gravity and there is greater danger of losing balance or straining muscles.

carrying objects

When carrying any object, the basic aim is to keep the center of gravity above the base of support and the basic alignment of body parts as close as possible to the normal standing or walking position. This is best done by holding the object as close to the body as possible and leaning away from the weighted side, being careful not to overcompensate. When carrying a package or object in front of the body, bending the elbows will provide a better angle of muscle pull. Extending the opposite arm slightly outward when carrying a bookcase or suitcase at the side will also help to maintain balance and keep the spine straight.

pushing and pulling

Many times in attempting to move heavy objects such as furniture and heavy appliances, it is necessary to push or pull in order to get them in position for lifting and carrying. Even then, the object may be too heavy for carrying without adequate assistance, and pushing or pulling may be the primary way to move it. By using proper techniques, the job can be made easier, back strain avoided, and the danger of injury from falling or slipping can be minimized.

Unless a great deal of sliding friction is involved, always apply the force at the object's center of gravity level. Applying the force too high or too low can waste energy. There is also the danger of tipping the object over if the force is applied too high. If sliding resistance is strong, the force is best applied below the center of gravity.

In getting set for pushing, spread the feet about shoulder width and place them in an up-back position, with the front foot close to the object. Bend the knees, keep the hips low, and the trunk inclined forward with the head up and back straight. Position the hands on a line with the shoulders. The force for the push then comes from extending the legs and not from pushing with the arms or back.

The same basic rules for positioning and application of force apply to pulling an object except, of course, that the hands are in a grasping position. Apply the force by leaning the body backward and downward, as if attempting to sit down, and slide the feet back quickly and alternately as the object is moved. With extremely heavy objects such as a refrigerator, moving one end at a time toward the rear foot and changing positions after each move is better than attempting a continuous movement.

postural defects and their correction

Knowing what good standing posture is, being "posture conscious," and keeping all opposing muscle groups in equal tone and length through a well-balanced exercise program may be all that is necessary for most people to avoid or correct certain postural deviations. For others, more specific remedial exercises or referral to an orthopedic doctor may be necessary to correct certain deviations which have developed over the years and to which the body has adapted.

The exercises that follow are for the various deviations shown in the Posture Analysis Form in the Appendix and are based on the principle of "Stretch the Short Side and Strengthen the Long Side." In cases where referral to an orthopedic surgeon or physical therapist is advisable, it will be indicated. Some of the exercises can be used as a supplement to a regular maintenance program for those whose normal work or regular sports activity require assuming positions or performing actions that can lead to muscular imbalance.

Unless indicated otherwise, the exercises should be performed slowly and with a great deal of concentration. Perform them daily, starting with the minimum number of repetitions shown and gradually increasing to the maximum. Apply appropriate overload techniques where possible. A page is provided at the end of the chapter for listing the specific deviations noted on your Posture Evaluation Form and the specific exercises for their correction.

lateral head deviation

One of the minor deviations, but one that can be unsightly is a head tilt to the side. Exercises for it are:

1. *Starting Position:* Stand or sit comfortably. *Action:* Without moving the shoulders, slowly move the head laterally towards the long side as far as

possible, as if attempting to touch the ear to the shoulder. Hold for six seconds, relax, and return to the starting position (10–20 reps.).

2. *Starting Position:* Same, but place the hand on the long side at the side of the head about ear level. *Action:* Simultaneously press with the hand and move the head towards the long side as far as possible. Hold for six seconds, relax, and return (5–10 reps.).

3. *Isometric Exercise for the Neck:* See Exercise 3 on p. 123. Place the hand on the long side.

low shoulder

If combined with a low hip, an anatomical deformity may be indicated and arrangements should be made for an examination and a possible shoe lift. Exercises include:

1. *Starting Position:* Stand erect with the arm of the low shoulder raised directly overhead. *Action:* Slowly bend to the long side as far as possible, hold for six seconds and return (10–20 reps.).

2. *Starting Position:* Stand erect, arms at sides. *Action:* Bend laterally to the long side as far as possible and return. Repeat in a moderate cadence from 10 to 20 times. If available, hold a weighted object in the opposite hand. (See Side Bend Weight Training Exercise 12, p. 121.)

3. If a chinning bar is available, hang with the arm of the short side (overhand grasp) for as long as possible, not exceeding one minute. Repeat 5 times.

scoliosis

Scoliosis is a lateral curvature of the spine, as seen from the rear. It may be simple or compound. In *simple scoliosis,* there is a single curvature referred to as a "C" curve, "right" or "left," away from the open side. For example, the normal "C" would be a curve left. Definite double curvatures are called "S" curves. An "S" curve is usually a compensatory curve.

The cause of a scoliosis is usually a low shoulder or low hip which causes a lateral tilt of the pelvis. In both cases, a shortened limb may be responsible or it may simply be the result of adaptation to faulty standing habits. A curve which deviates from a straight line by up to 1/2 inch is considered slight; by 3/4 inch moderate; and by 1 inch or more severe. If moderate or severe, corrective efforts should be made under professional supervision and only after orthopedic examination. If slight, it may be corrected with proper individual exercise. The more rigid the scoliosis, the more difficult it will be to correct. Rigidity may be tested by bending over in a hip flex position with the arms hanging loosely or hanging from a bar with two hands. If the spine straightens, you can be assured of good results, but even in rigid cases, improvement is possible. Exercises for simple scoliosis include:

1. Same as Exercises 1 and 2 for a low shoulder p. 152. For a "C" curve left, bend left; a "C" right (reverse "C"), bend right.
2. Sustained two-arm hanging from a bar with an overhand grip or sustained stretching from a back lying position with the feet hooked and stabilized and the arms stretched back above the head and pulling on a bar.
3. *Starting Position:* Sit on the floor with the knees bent and slightly apart. Reach forward with the palms facing outward, firmly grasp the feet, and tuck the head between the knees. *Action:* Keeping a firm grasp, slowly extend the feet so that you feel the pull along the spine, keeping the head well tucked. Hold for six seconds. Relax and repeat (5–10 reps.). See also: General Calisthenic Exercises 5 and 6, p. 110.

low hip

Since there is a possibility of a short limb, an orthopedic examination should first be made for possible prescription of a shoe lift. In some cases, the cause may be a low shoulder condition and may be eliminated by its correction. Otherwise, exercises that concentrate on restoring pelvic balance should be performed.

1. *Starting Position:* Stand erect, arms stretched and hands clasped overhead. *Action:* Slowly bend to the side of the low hip and return (10–20 reps.). (*Note:* Action contracts low side and raises pelvis).
2. *Starting Position:* Stand erect, hands at sides. *Action:* Slowly bend laterally to the side away from the low hip and touch the outside of the foot. You will have to bend the knee on that side slightly to touch the foot, but keep the other leg locked at the knee (10–20 reps.).
3. *Starting Position:* Stand erect with hands on the hips. *Action:* Slowly perform a deep knee bend, coming up on the balls of the feet to keep the trunk erect. Do not go beyond a right angle at the knee. Return and repeat (10–20 reps.).

pronated ankles

The height of the arch is not important to the function of the foot, but the mechanical balance of the bones of the feet to that of the body is. With the feet parallel and approximately 4 inches apart, a straight line should run from the knee cap, through the center of the ankle, and to the second toe. From the rear, a straight line should pass through the center of the Achilles tendon. When the ankles roll in away from this straight line, mechanical balance and good foot function is impaired. Pain and discomfort will ultimately appear. Exercises include:

1. *Starting Position:* Stand erect, feet 4 inches apart and parallel. *Action:* Curl the toes and shift the weight to the outside of the feet. Hold six seconds, relax, and repeat (10–20 reps.).
2. *Starting Position:* Same. *Action:* By the numbers (1–2 count), roll out on the outside of the feet and return (10–20 reps.).
3. *Starting Position:* Same. *Action:* By the numbers, rise up on the toes and return (10–20 reps.).

flat feet

Though there may be some relationship between the height of the arch and proper leg-ankle-foot alignment, good alignment may still be possible with the arch high, moderate, or even low. For many years the American public has been led to believe that flat feet, per se, is a serious disorder and eventually will lead to serious complications if allowed to go unchecked and untreated. This unfortunate myth probably began when thousands of American men were declared unfit for military service during World War I because of flat feet.

Just about all normal children have flat feet in their formative years only to develop strong arches as they grow older. Moreover, it is not uncommon to see many people with flat feet who do not suffer during normal routines of walking, running, or other activities. Many Olympic athletes have had flat feet and suffered no ill effects. Paavo Murmi, one of the greatest distance runners of all time, showed the world that flat feet are not a handicap, but a condition under which normal or better performance can be expected.

Only a relatively few people experience any pain with fallen arches and they, of course, need medical advice and relief. If proper alignment is present, the others should not be concerned and should go on their way without running to the corner drugstore for useless and unnecessary arch supports and other paraphernalia. It is improper alignment and footwear—including high heeled shoes—that cause most of our foot problems.

forward head

Forward head is another minor, but unsightly deviation which can be easily corrected. It is often associated with *kyphosis,* an excessive curvature of the upper back, coupled with forward shoulders and protruding or "winged scapulae" (shoulder blades). In such cases, the exercises for these conditions may be used. For a "poke neck" deviation alone, the following exercises can be performed:

1. *Starting Position:* Stand or sit comfortably. *Action:* Slowly lower the head backward as far as possible and hold for six seconds. Return and repeat (10–20 reps.).
2. *Starting Position:* Same, but clasp the hands behind the head. *Action:* Simultaneously press with the hands and move the head slowly back as far as possible. Hold for six seconds, return, and repeat (5–10 reps.).
3. Isometric Exercise for the Neck. See Exercise 2, p. 123.

kyphosis (round shoulders)

Kyphosis, an excessive curvature of the upper spine, may sometimes be caused by a short clavicle, or collarbone, which tends to round the shoulders. In such cases, correction is limited. Most conditions, however, are the result of squeezed chest muscles and overstretched back muscles that pull the scapulae toward the spine, caused by adaptation to faulty positions. It is usually combined with a forward head condition.

As in scoliosis, the condition may be slight, moderate, or severe and the rigidity of the condition will determine the ease of correction. The degree of rigidity can be determined by bending the head and back over to maximum extension (with assistance, of course) and seeing and feeling if the curve straightens. In rigid or severe cases, professional advice should be sought. Otherwise, the following exercises can be performed, along with the exercises for forward head. Where appropriate and available, the use of dumbbells can be made to increase resistance.

1. *Starting Position:* Stand at attention, tall and erect. *Action:* Slowly swing both arms forward and upward, reaching overhead to a full stretch and at the same time rise high on the toes. Now turn the palms outward and lower the arms sideward and downward while pressing them back forcefully. At the same time, pull the chin in, keeping the head high, and let the heels drop to the ground. Be careful to avoid excess arching of the lower back (10–20 reps.).
2. *Starting Position:* Stand with the feet about 6 inches apart, knees slightly bent, and bend forward at the hips to a 90 degree angle. Relax the trunk and neck, allowing the arms and head to hang loosely. *Action:* Swing the arms sideward and upward vigorously. Bring the head up, retract the chin forcefully, and flatten the upper back. Hold this position for two seconds and return (10–20 reps.). (See also Weight Training Exercises: Bent-Over Lateral Raises, p. 121).
3. *Starting Position:* Stand tall and erect with the fingers touching the tips of the shoulders, the arms in front of the chest, and the elbows pointing downward. *Action:* With the elbows hugging the sides, slowly move the

arms outward and back as far as you can go. Hold for six seconds while attempting to retract the chin and stretch the body upward. Return and repeat (10–20 reps.). See also: Warm-Up Exercise 5, p. 108, Weight Training, Bent-Over Rowing Exercise, p. 121; Isometric Exercise C-1 under *Back, Abdominal, and Gluteal*, p. 124).

trunk hyperextension

Trunk hyperextension is often associated with a sagging abdomen and *lordosis*, the technical term for an excessive curvature of the lower back. Even as a sole condition, exercises should be aimed at strengthening and shortening the stomach muscles and stretching the lower back and hamstrings. These exercises are described below under Sagging Abdomen and Lordosis.

sagging abdomen

Two of the best exercises for weak abdominal muscles are the Bent-Knee Sit-Up and the "V" Sit-Up:

1. *Bent-Knee Sit-Up:* See Weight Training Exercise 10, p. 121.
2. *"V" Sit-Up: Starting Position:* Back lying, legs straight, feet together, arms at sides with the palms down. *Action:* Simultaneously raise the trunk and legs to about a 45 degree angle, moving and extending your arms forward at the same time to maintain balance. Return (10–20 reps.).
3. *Gather Sit-Ups:* See Calisthenics Exercise 7, p. 110.

lordosis

Lordosis is an excessive curvature of the spine at the lower back and is commonly referred to as "swayback." In some persons, the condition is congenital due to placement of the sacrum (lower portion of the spine) at a greater angle in the pelvis. (See Fig. 10.) Correction is limited in these cases. But the muscles of the abdominals, low back, thighs, and hamstrings also play a key role in controlling the position of the pelvis. Deficiencies in these muscles often compound the problem for persons with congenital weakness and is the principle cause of swayback in others.

The abdominals play a major role. If these become weak and stretched, the pelvis is tilted forward and the lower back curve is increased. Tightness and shortening of the thigh, hamstring, and low back muscles also cause the pelvis to slant forward. Slackness in the gluteals also plays a part, since good muscle tone here helps to lift the pelvis upward.

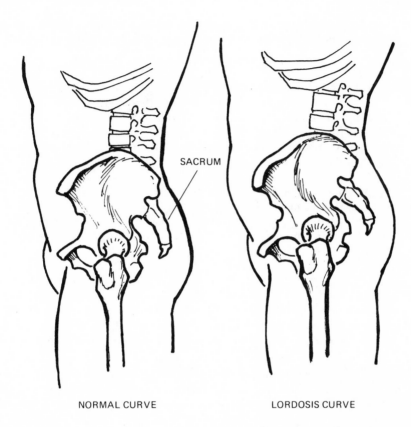

NORMAL CURVE   LORDOSIS CURVE

Fig. 10.   Lordosis Curvature

Exercises for the alleviation or correction of a swayback condition, then, include those which strengthen the abdominal and gluteals and those that stretch the low back, thigh, and hamstring muscles. For the abdominals, the Bent-Knee Sit-Ups and "V" Sit-Ups are best. Straight Leg Sit-Ups should be avoided since these tend to contract the low back and front thigh muscles which can only aggravate the condition. The following exercises are recommended in addition to the abdominal exercises:

1. *Back Flattener or "The Bumps": Starting Position:* Back lying, with legs bent and arms at the sides, palms down. *Action:* Simultaneously contract the abdominals and gluteals so that your back is flattened against the floor. Hold for six seconds, relax, and repeat (10–20 reps.). As you progress, the same exercise can be performed with the legs straight. Contracting these muscles often during the day while sitting and concentrat-

ing on keeping the gluteals contracted while standing or walking will also help to keep these muscles in good tone and the pelvis up.

2. *Back Stretch: Starting Position:* Same. *Action:* Contract the stomach and flatten the back as in the previous exercise, then raise the head, bring both knees to the chest, clasp both hands around the knees, pull tightly and hold for six seconds. Relax and return (10–20 reps.).

3. *Alternate Hamstring Stretch: Starting Position:* Stand erect, feet together, arms at the sides. Bend the legs, reach down, and grasp the ankles. *Action:* Alternately and at a moderate rhythm, straighten each leg by locking the knee. For those with very tight hamstrings, it might be necessary to grasp a little higher, As the movement becomes easier grasp further down, attempting ultimately to be able to straighten each leg with the fingers or knuckles touching the floor. See also Warm-Up Exercise 8, Knee Lifts, p. 109; Calisthenic Conditioning Exercise 1, Toe Touch, p. 110; and Exercise 5, Sitting Stretch-Static, p. 110.

low back pain

Low back pain is one of today's most common complaints among middle-aged and some young people. It is conservatively estimated that 30 million Americans, nearly one in every six people, have sought relief from this nagging ailment. In the overwhelming majority of cases, the causes may be traced to faulty body mechanics, weak or tense muscles, or a combination of these factors. This places the problem largely in the category of being unnecessary and avoidable.

faulty body mechanics

Faulty body mechanics may be responsible for low back pain by causing an excessive compression of the lower vertebrae, allowing them to actually press together, or by causing direct muscle or ligament strain. Normally, the vertebrae are separated by elastic discs of cartilage which allow flexibility of the spine and serve as a cushion to protect the underlying nerves. Poor posture positions which allow the pelvis to slant downward and excessively curve the lower back may flatten these discs to the point that the pressure on the nerves is severe enough to cause pain. These positions may also cause a disc to rupture or slide forward to the point that the vertebrae come into direct contact with each other, irritating the nerves and causing pain. In severe cases, surgery may be necessary for correction. Improper lifting, pushing, or pulling with the relatively weaker lower back muscles rather than the stronger leg muscles are often the cause of direct muscle and ligament strain. By following the principles of good body mechanics as outlined earlier and correcting muscle weakness that may be the cause of a swayback condition and excessive compression of the vertebrae, at least one major cause of low back pain can be avoided.

muscle weakness and tenseness

Problems caused by compression of the vertebrae, however, are only 20 percent of all low back cases. According to back expert Dr. Hans Kraus of the New York University College of Medicine, 80 percent of all low back problems are the result of simple muscle weakness or tightness in one or more of the key muscle groups that support and control movement of the torso. In addition to the back muscles themselves, these include the abdominals, the hip flexors (psoas group), hip extensors (gluteals), and the hamstrings. It is lack of strength and flexibility in these muscles, caused by inadequate physical activity, which predisposes the back to both major and minor strains.

Dr. Kraus's conclusion was based on clinical data involving over 5,000 chronic and acute back sufferers. Examination of these patients showed that in 80 percent of the cases, there was no evidence of any "specific disease, lesion, or anomaly (disarrangement) which could be blamed for their affliction."[1] Muscle weakness and stiffness were the sole causes and with their reduction by proper strengthening and stretching exercises, the condition of 91 percent of a control group improved from fair to good. Back pain caused by muscle weakness and tightness is referred to as the *low back syndrome* and is characteristically found among sedentary persons who develop the condition from long periods of sitting—combined with inadequate exercise of the key muscle groups mentioned.

Prolonged sitting with the hips and knees flexed keeps the psoas and hamstring muscles shortened and in a tense position. The muscles of the back are also contracted in a static position, while the stomach muscles sag. It is the constant tensing of these muscles, without adequate release, that plays a major role in the low back syndrome. If tensed frequently without being allowed to return to its normal, full-length position, a muscle will lose its elasticity and shorten permanently. Constant tensing of these key muscle groups, then, tends to tighten and shorten them. To the weakness caused by disuse, another is added. In the back and hamstring muscles, especially, if the shortened muscle is further contracted by any additional stimuli (nervous tension or any

---

[1] Hans Kraus, M.D. and Wilhem Raab, M.D., *Hypokinetic Disease* (Springfield, Illinois: Charles C. Thomas, Publisher, 1961), p. 17.

physical use), the additional muscle tension can lead to contracture, spasm, and pain. The pain may be localized in the muscle itself or radiated to other parts of the body via the nervous system. Tightness in the hamstring muscles, for example, often causes referred pain in the lower back. Similarly, pain in the lower back can be radiated to the neck, shoulder, or extremities. The referred pain often leads to a curtailment of activity and additional weakness. It is this combination of muscle tightness and weakness, according to Kraus, that sets the stage for the first episode of back pain. "Picking up a paper or pencil may precipitate the first attack. It leaves the muscles weakened and more stiffened—ready for the next episode of pain which in turn will compound the symptoms."[2]

avoiding and relieving low back syndrome

The key to avoiding low back syndrome is obviously keeping the key trunk support and movement muscle groups strong and flexible through appropriate exercise. Developing and maintaining strength of all major muscle groups through one or more of the general exercise programs presented and other physical activities, games, and sports appears to be the major preventative. Such a regular program of exercise has a double value in serving as an outlet for nervous irritations and tension that can contribute to muscle tightness.

For those with specific postural defects (lordosis), addition of appropriate corrective exercises should be made. The following exercises are designed specifically for those who are actually suffering from low back pain caused by muscle weakness, strain, or tension. If there is any question of structural damage, of course, a physician should be seen first. The exercises are aimed at regaining minimal strength and flexibility of the key muscle groups involved in low back pain. They should be considered corrective, developmental; they are a prelude to a more strenuous and well-rounded program, but can be included in a regular maintenance program for those who may be prone to back problems.

remedial exercises

All exercises should be performed on a firm surface, with from 1 to 20 repetitions, depending on individual condition and tolerance level. Perform them slowly and allow for complete relaxation between movements.

2 Kraus, *Hypokinetic Disease*, p. 19.

1. *Preparation: Starting Position:* Back lying, arms at sides, palms down, knees flexed with a pillow or by placing the feet flat on the ground. *Action:* Breath slowly and deeply through the nose and exhale slowly through pursed lips. Attempt to relax all the muscles before starting the exercises (5 times).

2. *Back Flattener in Bent-Knee Position.* See Lordosis Exercise 1, p. 157.

3. *Back Stretch:* See Lordosis Exercise 2, p. 158.

4. *Psoas Strengthener and Back Stretch: Starting Position:* Back lying, arms at sides, legs extended, with knees relaxed. *Action:* Contract the stomach and flatten the back; then raise the head and bring one knee to the chest as far as possible. Keep the other knee relaxed. Return to the starting position and repeat with the other leg.

5. *Psoas Stretch and Gluteal Strengthener: Starting Position:* Front kneeling position with weight supported on hands and knees, looking down with the head, neck, and back straight. *Action:* Keeping the head and back straight, raise one knee up and back as far as possible while keeping the leg flexed. Be careful not to arch the back. Return to the starting position and repeat with the other leg.

6. *Low Back Strengthener and Hamstring Stretch: Starting Position:* Standing, with feet together and hands at sides. *Action:* Slowly bend forward at the hips and at the same time extend the arms backwards. Return slowly to the starting position.

7. *Sitting Stretch-Static:* See Calisthenic Conditioning Exercise 5, p. 110.

8. *Alternate Hamstring Stretch:* See Lordosis Exercise No. 3, p. 158.

9. *Modified Bent-Knee Sit-Up:* Starting position with hands at sides. For some conditions, this may have to be delayed.

general tips for low back care

1. Avoid prolonged standing in one position. It is difficult to keep the hips from sagging forward after standing for a long time and this places a strain on the lower back. To relieve the strain while doing various tasks (working at a table, washing, ironing, etc.), flex one of the hips by placing a foot on a stool, step or railing.

2. Always bend the knees when leaning forward, lifting, or lowering any object.

3. Avoid carrying objects above the level of the elbows.

4. Never sleep on your stomach without a pillow support under it.

5. Avoid lying flat on your back without flexing the knees. When resting, place a pillow under the knees to keep the back flat.

6. When sitting, keep the knees parallel or slightly higher than the hips. If driving, be sure the seat is up close to the pedals so the legs do not have to reach out.

7. Avoid positions in which the neck and head are thrust forward with the chin tilted up.

personal posture correction guide

Deviation   *Corrective Exercises*

Notes

7

weight control through diet and exercise

A major component of personal fitness is maintaining proper body weight and proportions. While being overweight or obese is the most serious problem, other causes of concern also exist, including underweight and fluctuating weight. In any case, permanent weight control depends on a continual adjustment of diet to levels of activity and the development of sound eating habits.

The purpose of this chapter is to provide the understanding and tools necessary for proper weight control through diet and exercise. Included will be a discussion of some fundamental principles and a specific method of structuring a sound diet and exercise program to bring about desired changes. First, a brief review of the three major weight problems and their dangers.

weight problems and dangers

overweight

The magnitude of the overweight problem in our nation and its relation to cardiovascular disease was pointed out in Chapter 1. Not only is overweight one of the major single risk factors in the incidence of heart attacks, it is also related to the development of high blood pressure, high levels of cholesterol and fatty acids in the blood, and diabetes—other major risk factors in cardiovascular disease.

A primary reason for the close association between overweight and heart attacks is the increased strain that excess body weight places on the heart and circulatory system. An increase in body fat increases the body surface area, requiring an expansion of arteries, veins, and capillaries in order to serve the larger area. It is often said that "20 pounds of fat means 20 miles of capillaries." This, of course, places a much greater work load on the heart, forcing it to continually pump blood through a greatly expanded vascular system. Extra stored energy, or fat, also tends to accumulate in the connective tissue beneath the membrane that encloses the heart. As the additional fat becomes extensive, the mechanical action of the heart becomes impaired.

Overweight as one of the major contributors to high blood pressure has been demonstrated by researchers. High blood pressure has been found to be more than twice as common among the obese than those of average weight. Fat in the form of cholesterol or triglycerides accumulates in the blood vessels and reduces the size of the arteries. The increased resistance to blood flow raises both systolic and diastolic blood pressure and places an added strain on the heart. The same narrowing of the arteries from fat deposit accumulations (atherosclerosis) is a primary cause of coronary heart attacks, strokes, and crippling disability in the legs and arms.

Overweight persons face additional dangers from increased susceptibility to kidney and liver ailments. They are also a surgeon's nightmare, considered to be poor surgical risks because of the difficulty that excess fat causes in operating procedures. Recovery time following accident, illness, or surgery has also been found to be slower, though this is probably more from the general lack of conditioning found among the obese caused by inactivity, a major factor in their excess weight.

underweight problems

Generally speaking, the problems faced by the underweight person do not reach the proportions of the overweight or obese individual. However, they are not completely free of danger. The most serious health danger is a greater susceptibility to respiratory and nervous ailments. In addition, they are generally hampered by a lack of muscular development and strength, some lack of endurance, less vitality, and increased susceptibility to infections. In general, being more than 10 percent below average on standard weight tables or having less than 10 percent body fat may be considered an indication of underweight.

Being underweight is not to be confused with malnutrition which is a broader term covering nutrition in relation to body needs.

fluctuating weight

Fluctuating weight may signal a disturbance in the endocrine system controlling metabolism brought about by any number of minor or major internal disorders or illnesses. For this reason, any drastic change in weight for no apparent reason calls for a medical consultation and examination. Many persons, however, go through periods of excessive weight gain and loss from a failure to establish a consistent pattern of eating and exercise. These are the people who when they notice their belts or dresses becoming tighter go on a "crash diet," possibly supplemented by exercise. But when they get down to their desired weight, they quickly fall back to the same patterns that caused the original weight gain and are soon repeating the cycle.

The practice is a poor one for several reasons. Because of their continual opening and closing, the fat cells become larger, stronger, and more resistant to closing. This makes them more receptive to fat deposits. Weight control, then, becomes much more difficult. The constant fluctuation also contributes to an unstable metabolism and creates a greater chance of "fouling up" the body's complex metabolic mechanism, possibly leading to other serious problems.

fundamental principles and facts about weight control

Many factors must be examined in learning to control body weight effectively and safely. Before going into the specific method of establishing a weight control program, certain factors should be discussed so that the discussion is firmly based on understanding.

calorie intake and expenditure— the essence of weight control

Several years ago, a notion became popular that "calories don't count" when it comes to controlling weight. The notion gave false hope to many overweight persons looking for an effortless way to lose weight without bothering to watch their eating habits. Though admitting there are factors that affect the nature and efficiency of calorie burn-off—including the proportion of protein, carbohydrates, and fat in the calorie intake— the overwhelming majority of nutritional experts disclaimed this false notion. They restate the basic principle: "Weight loss and weight gain depend essentially on the ratio between calorie intake and calorie expenditure."

As we have seen, all body activity requires energy produced by the oxidation of food and oxygen. A *calorie* is a unit of measurement used to describe the energy produced by food when oxidized in the body. Specifically, it is the amount of heat required to raise the temperature of one gram of water from 14.5 to 15.5 degrees Centigrade. Food energy taken in and not used by the body for its activities is stored as fat for later breakdown and use as needed. For example, during the course of a day, one might take in a food value of 3,000 calories. Yet, the level of body activity may be such that only 2,500 calories are burned off. The surplus of 500 calories is then stored throughout the body, to be drawn upon at any time the calorie intake falls below the amount needed for a particular level of activity. Since one pound of fat is equivalent to 3,500 calories, repeat this pattern over a period of time and the result will be excess fat and body weight at the rate of one pound every seven days. The reverse of excess stored energy is seen when calorie burn-off exceeds the intake and weight loss results.

It should be remembered that, in terms of quantity, it takes five times more body fat than lean muscle mass to equal one pound of fat. This means two things: (1) Though the extra pound on the scale may not seem like much, it is significant in terms of the increase in body fat percentage. (2) When combining an exercise and diet plan or just using an exercise program to control weight, your "scale weight" may indicate little or no progress; but if measured in terms of body fat percentage (skinfold caliper measurements), a considerable drop may be noted. It would not be uncommon to maintain or actually gain "weight" over a three month, four day a week program and actually drop over 5 percent in body fat percentage. Indeed, a forty-five-year-old woman in one of our classes dropped from a size 14 to a size 10 dress in nine months without losing a single pound! Others have shown equally striking results.

At any rate, it is obvious that to maintain a given body weight and body fat percentage a balance must be struck between the intake of energy from food and its expenditure through various activities—including basal metabolism (vital functions when the body is at rest), normal daily activities, and extraordinary activities. Total calorie needs will vary with each individual. They may range from as little as 1,200 for a woman doing sedantary work to 6,000 for a young man practicing for the pentathalon or a similar vigorous sports activity. As noted, however, there are factors which may and do affect the rate and efficiency of energy breakdown and conversion for these activities and cause variations among persons taking in an equivalent number of calories and engaging in essentially the same type of activities.

factors complicating weight control[1]

There are a number of factors that complicate the relatively simple principle of controlling weight through balancing calorie intake and expenditure. These work to greatly modify individual intake and need for calories and it is important to recognize them if a proper understanding and control of body weight is to be gained. In some cases, professional medical assistance may be necessary for proper evaluation should the adjustment of diet and exercise levels fail to produce effective results.

Digestive system efficiency. There is a definite variability in the digestive systems of individuals and their ability to break down the energy component of food. A person with an efficient system is able to supply the body with more calories from the same amount of food than one with an inefficient system. This creates the need for a greater expenditure of energy through activity or a lower intake for an equal amount of work. Digestive system efficiency, however, is difficult to determine without extensive medical tests.

Basal metabolic rate. The basal metabolic rate refers to the rate at which the body uses energy just to maintain itself during a state of complete rest. When medically determined, it is taken after a normal night's sleep and before breakfast by measuring the amount of oxygen used by the body for ten to fifteen minutes. From this, basal heat production is calculated on the basis of 4.82 calories of heat for every liter of oxygen used.[2]

Normally, basal metabolism is directly proportional to total body surface area which, in turn, is related to the height and weight of an individual. After age five, the BMR for males is about 5 to 10 percent higher than for females. For both sexes, the rate drops markedly at age twenty and continues to decrease slowly with age.[3] Studies measuring basal metabolism on the basis of active muscle tissue alone, however, have failed to show such a decrease, leading some researchers to wonder if the decreased rate normally attributed to age is merely a reflection of

1 W. C. Adams et al., *Foundations of Physical Activity* (2nd ed.) (Champaign, Illinois: The Stipes Publishing Company, 1965), pp. 96–97.

2 Wayne Van Huss, et al., *Physical Activity in Modern Living* (2nd ed.) (Englewood Cliffs, N.J.: Prentice-Hall, Inc., 1969), pp. 142–44.

3 Van Huss, et al., *Physical Activity,* pp. 142–44.

fat accumulation caused by decreased physical activity and overeating. In any event, the basal metabolic rate may account for as much as 50 percent of a person's total energy requirements even in a physically active person, and an even larger percentage in one who is inactive. The point is that since it influences calorie expenditure at all times and is such a large factor in determining energy needs, a slight difference or change in the rate can significantly change a person's overall calorie needs.

The body's regulator for the basal metabolic rate is thyroxin, a hormone secreted by the thyroid gland. If there is a deficiency in the amount produced, a lower metabolic rate will result, thus reducing the total calorie requirement. Conversely, an overproduction of thyroxin will increase metabolic activity and calorie expenditure. Malfunctioning of the thyroid and other glands, however, appears to play only a minor role in the problem of overweight. In one study of 275 obese individuals, less than 3 percent had a glandular disorder that could be blamed for their problem.

Heredity and environment. Studies have shown that three out of four obese individuals come from families with a history of overweight. Whether this tendency toward overweight is caused by hereditary factors or by acquired family eating habits, however, has not yet been clearly determined. Heredity is a possible factor, since it has been shown to have a definite influence on the functioning of the endocrine glands. But the more likely cause seems to be the developmnt of the overeating habit from family eating habits and culturally instilled attitudes. In some families, preparing and serving a continuing array of attractive, usually high calorie meals is considered more an expression of love rather than a means of providing needed nourishment for the body. Though this is not to deny the psychological and emotional values of eating, going to the extreme of making every meal a feast is far from a true concern for the welfare of a loved one. Another culturally developed attitude that can lead to overeating is one that insists "the plate must be left emptied." Overeating habits developed from such attitudes and patterns instilled from early childhood are difficult to break.

Emotional overeating. To some people food becomes an escape mechanism for gaining satisfactions and security otherwise missing in their lives or during times of emotional stress. When lonely, frustrated, bored, or unhappy, they turn to food for psychological release. In many respects, it is similar to alcohol, smoking, or drug addiction. This tendency is also often developed from early family practices of using food as a "reward" for good behavior or as a means of consolation during times of illness or other difficult times. In many overweight persons the problem is basically finding the psychological problems that are causing the person to overeat.

Water retention. The intake of water is reliably regulated by the sensation of thirst. Water loss, however, is related to total metabolism. In some persons, particularly women, a distressing factor in efforts to control weight is a tendency toward excessive water retention. The condition is called *edema* or bloating, and has nothing to do with calorie intake or expenditure. Edema in women quite often begins just before the menstrual period and continues throughout the period. Edema during such times is normally a temporary condition. During such times or for those with a more persistent problem, the elimination of certain foods such as salt, pickles, salted butter, salted or cured fish and meats, and crackers can be helpful. Recently it has been found that vitamin C is effective for the temporary relief of edema. When necessary for those with special water retention problems, a variety of diuretics (drugs which increase urinal production) may be prescribed by a physician. Normal variances in the degree of water retention from various foods may also cause body weight to vary slightly from day to day, influenced also by atmospheric temperature, humidity, and levels of physical activity. For this reason when using a daily scale weight check to keep body weight stable, it should be done just after rising and before breakfast, and an allowance of from one to three pounds made for daily variance.

Other factors. Other factors complicating the calorie intake-expenditure principle of weight control are an individual's proportion of lean and fat body mass and the type of nutrients in the calorie intake. Fat is generated from food at the membranes of the fat cells whereas lean cells convert fat and other nutrients into energy by oxidation. Fat people with their greater number of fat cells, therefore, have a greater capacity for storing fat than lean people. Conversely, the more lean mass area, the more energy burn-off before the excess is converted to fat. The significance lies in knowing the priorities and ways in which the body breaks down food nutrients into energy (which varies according to the proportion of nutrients and type of activity) and in the type of efforts that must be made to increase and maintain lean body mass through diet and exercise.

diet and weight control

Every year seems to usher in a new series of "fad diets" claiming to help a person lose weight effortlessly and painlessly. Most of these quick reducing diets lack the proper balance of essential foods, and a person may do great harm to himself by practicing them. Furthermore, they seldom result in any permanent weight control because they fail to bring about a basic change in eating habits. Unless diets are considered from a long-range view and can contribute to the adoption of a daily, lifetime style of eating that a person can live with, they can be of only limited value.

There are also many eating practices which are based on misconceptions and poor nutritional structure. These include vegetarian diets, avoiding certain food combinations, or eating certain foods in the belief they will cure various disorders.

The vegetarian diet is based on the theory that all animal products should either be avoided by humans or at least limited to eggs, milk, and cheese. There is no scientific evidence to support this belief. In fact, the human digestive system (as opposed to the herbivorous or plant-eating animals) is specifically designed for the digestion of both vegetables and meat, and functions best when both are included in the diet.

Another false belief is that mixing certain foods can be harmful—such as acid and alkaline food products (cucumbers and milk). Again, there is no evidence to support this view. Neither is there any that taking or avoiding certain foods will cure disorders, such as that citrus fruit will cure a cold or that avoiding meat will arrest high blood pressure. With the exception of medically structured diets for persons with certain conditions, the well-founded and balanced diet will offer the greatest nutritional protection.

basic methods of dieting

It should be evident that weight loss by dieting can be accomplished by calorie reduction only. There are two basic methods that can be used: (1) eat the same foods you are accustomed to, but cut down the size of the portions; or (2) restrict food intake by actually counting calories taken in. In using the second method, two precautions must be taken: (1) an adequate food calorie chart or predetermined total calorie meal plan should be used; and (2) You must be certain that the diet adequately meets the nutritional requirements of the body.

nutritional requirements

Nutrition is the series of processes by which an organism takes in and assimilates food both for the growth, repair, and maintenance of tissues and in order to meet its total energy requirements. The three major classes of nutrients are carbohydrates, protein, and fats. Other nutrients include vitamins and minerals. Each of these nutrients serves both a general and specialized purpose.

Carbohydrates. Though carbohydrates, proteins, and fats all serve as energy sources, the great energy-producing elements and the primary source of fuel for physical activity are carbohydrates. For less strenuous activities, a mixture of carbohydrate, fat, and protein is oxidized for food, but as activity becomes more strenuous, the body prefers carbohydrates.

Carbohydrates include sugars, starches, cellulose, and glucose. The simplest is glucose, found naturally in fruits and honey. Other sugars and starches are broken down into glucose by the body's complex digestive system before they can be utilized. Glucose or simple sugar can be directly oxidized for immediate energy, converted into glycogen and stored in the liver or muscle tissues or, if in excess, converted into stored fat. The body, however, will always store glucose as fat before its capacity to store glucose as glycogen has been exhausted. It is this continuous storage of excess carbohydrates, of course, that leads to obesity.

Some glucose is continually carried in the blood stream (called blood sugar) and is necessary for the body's various metabolic activities and for all muscular contractions. The liver helps to keep the blood sugar level at a normal operating level by balancing the amount of glucose it secretes into the blood and the amount it takes from the blood and stores as glycogen. The amount which it can store, however, is not enough to maintain a proper operating level for long periods of time and is directly related to carbohydrate intake. When a low level of blood sugar occurs, a person often experiences weakness and irritability. If extremely low, there is also danger of convulsions. For these reasons, low carbohydrate diets are inadvisable unless medically prescribed.

Not all carbohydrates can be broken down by the human digestive system, and these are of limited use to man. An example is cellulose, found in the cell walls and woody parts of plants. The herbivorous animals are able to break down cellulose because of their extra stomachs and the use of certain bacteria and enzymes not inherent in man. For humans, the only value of such cellulose-loaded plants (lettuce, for example) is to provide roughage and flexibility in waste elimination.

Protein. Protein is the main substance of the human body and is present in all the muscles, organs, skin, hair, and tissues in various combinations of chemical elements. All protein molecules are made up of smaller units called *amino acids.* Some proteins have only a few amino acids while others have many. Actually, there are about twenty-five known amino acids, with some coming directly from food and others synthesized (made or built from simpler chemical elements) in the body. Each serves a particular body-building or repair function. From a safe dietary view, it is important to know that amino acids are not stored in the body and so should be included in each meal.

Though the primary function of protein is to provide the raw material for the growth and repair of tissues, it can and is also used to provide heat and energy. As stated previously, at low levels of activity, some protein is oxidized along with a mixture of fats and carbohydrates and used as a source of energy. In an emergency, of course, or when other energy sources are at a low level, greater amounts of protein are oxidized for energy use. It should be noted, however, that increasing protein intake in preparation for periods of vigorous activity has been shown to be of little value. Experiments by German scientists showed that during exercise no extra protein was metabolized above that in a resting state, indicating that exercise does not result in an increased rate of wear and tear of muscle tissues.[4] Other studies on dogs showed that even after long periods of vigorous activity, there was no increase in energy metabolism. The fairly common practice of special high protein diets for athletes during a playing season, and pregame meals of steak appear to have no other value than psychological. A more effective pregame meal plan would be an easily digestible high carbohydrate meal, timed to provide for maximum energy when most needed. In weight training or other exercise programs aimed at building up lean muscle mass, increased protein intake is recommended.

Fats. Measure for measure, fats provide more energy than either protein or carbohydrates. Chemically, they are compounds of fatty acids and glycerol (a sweet, syrupy alcohol) containing carbon, hydrogen, and oxygen—the same elements found in carbohydrates, but in different proportions. Fats contain more carbon and hydrogen, but far less oxygen. This makes them a more concentrated form of fuel, but at a far greater oxygen consumption cost for oxidation. For this reason, high fat diets tend to produce greater fatigue. For those engaging in endurance type activities, such diets are definitely not advisable.

Ingested fats are broken down by digestion into their fatty acid-glycerol components by a much more complicated process than carbohydrate digestion. Once absorbed into the bloodstream, they are used for transporting fat soluable vitamins (A, D, E, and K), as a source of energy, and for other purposes. Excess amounts are stored with other body fat deposits. The liver temporarily stores a large percentage of the absorbed fat by converting it to certain types of fatty acids which it then releases into the blood stream as needed. It also collects and

4 Van Huss, et al., *Physical Activity,* p. 148.

stores fat from the circulatory system in the same way and also synthesizes fat from nonfat sources, such as carbohydrates and possibly proteins. If something happens to its fatty acid conversion mechanism, the fat deposits accumulate with possible serious results.

Stored body fat is formed from fat, carbohydrate, and protein foods, but usually from fat and carbohydrates first. When not utilized in the process of building body tissue, amino acids may be converted into glucose or fat. In addition to serving as a reserve energy food, the body fat helps to pad, protect, and support tissues and organs and to act as an insulator against heat loss. When needed, the fat in the liver and muscle tissue storage depots is mobilized by an active enzyme, called *lipase*. (An *enzyme* is an accelerating agent for specialized chemical breakdowns in body metabolism.) The lipase acts to break the fat into its fatty acid-glycerol components so that it can be absorbed into the blood stream and delivered to the active tissues. The tissue lipase controls the fat mobilization in the blood stream in the same manner as the liver stores and regulates the release of sugar. Upon arrival at the active tissues, oxidation takes place through a series of steps which reduce the higher fatty acids to simpler ones. These are broken down further and finally oxidized to produce heat, energy, and the final end-products of carbon dioxide and water. It is because of this much more complicated process, which naturally requires more energy in itself, that high fat–low carbohydrate diets have been advanced as allowing a person to eat more and still lose weight.

It is believed, however, that the complete oxidation of the fats requires that sugar or carbohydrates be oxidized at the same time; so it is good to remember the saying: "Fats burn in the flame of carbohydrates." If carbohydrate intake is low and fat oxidation is left incomplete, ketone bodies are formed in the blood (acetic acid compounds). These are normally eliminated by the kidneys. But in high fat diets or when the body fat stores are being used entirely, such as under starvation conditions, a large amount of ketones are formed. The condition, known as *ketosis*, is also caused by fevers, diabetes, hyperthyroid conditions, pregnancy toxemas, and other illnesses. In essence, then, the body's metabolism begins to resemble that of persons with these conditions with their potential dangers to the heart, kidney, and other organs.

Minerals. Minerals furnish no energy, but play an important nutritional role. The special functions of minerals are diverse. Some are necessary for growth and repair of bones and teeth. Others are needed to assist enzyme action in the vital role of catalyzing specific transformation of materials in metabolic activity. Several concern themselves with the formation of body fluids and secretions.

Though twenty-five minerals have been found in the body, the exact use and function of all is still unknown. Of those known to make definite contributions, calcium is the most common—2 percent of the total body weight is composed of this mineral. It is found in bones, teeth, cells, and body fluids, and assists in controlling the blood level. Phosphorous is another mineral found in the teeth and bones and plays an important part in the functioning of proteins and enzymes. Iodine plays a major role in forming thyroxin. Fluorine helps to keep the bones and teeth strong and healthy. Iron is necessary for the production of hemoglobin, the cells which carry oxygen in the blood, along with copper which in minute quantities is used as an oxidizing enzyme in hemoglobin production. Other minerals that make known contributions to body functions are magnesium which is involved in neuromuscular activity, manganese which serves various enzyme systems, cobalt which is used in the formation of red corpuscles, and zinc which assists in breaking down bicarbonates.

Vitamins. Vitamins are essential to maintain life. They serve a special purpose by furnishing small amounts of chemical substances which influence such body activities as digestion, assimilation, tissue building, and energy release. Except for vitamin D, the body cannot manufacture vitamins. The usual and most reasonable source of vitamins is food. There are times, however, when commercially produced vitamins might be necessary as a food supplement. Such instances are when a person is not consuming an adequate diet because of ignorance or poverty or during illness, emotional stress, and when on a special medical or reducing diet. In these cases, vitamins may be added to the diet, but they should never replace it. The self-prescription of vitamin mixtures so commercially popular today is a potentially dangerous practice; they should be incorporated as part of the daily diet only upon advice of one's physician.

An excess of vitamins A and D is one of the potentially dangerous practices. Excessive amounts of vitamin A may cause lack of appetite, weight loss, irritability, and cracking and bleeding of the lips. Excessive amounts over a long period of time can cause liver enlargement and loss of hair. Though children appear to be more susceptible to vitamin A intoxication, everyone should avoid excessive intake. An excess of vitamin D may cause nausea and diarrhea. When prolonged, calcium deposits take place in soft tissues, weakness appears, and mental depression is common.

Much has been written in recent years about vitamin C or ascorbic acid and its wonder powers in preventing colds and assisting tissue growth. In response, many persons have taken to the practice of supplementing daily diets with large quantities of commercially produced vitamin C. Until further investigation substantiates the value of this practice, however, the best of current medical advice is that the necessary amount can be obtained from a normal diet that includes fruits and vegetables—especially the citrus fruits. Strawberries, cantaloupe, tomatoes, green peppers, broccoli, raw greens, cabbage, and white potatoes also contain sufficient quantities of vitamin C. Since it cannot be stored, it must be taken in daily. A lack of vitamin C leads to scurvy, a condition in which the blood capillaries are damaged and hemorrhages occur in the tissues. One of its first signs is bleeding of the gums. In addition to preventing scurvy, vitamin C sources can quickly alleviate the condition.

Some vitamins are fat soluble (A, D, E, K) and others water soluble. Those which are fat soluble (pass into solution with fat) can be stored and rationed in the body more easily. Water soluble vitamins are easily lost through cooking and food processing.

Fat soluble vitamin A is found mostly in yellow fruits, dark green and yellow vegetables, cream, butter, whole milk, cheddar-type cheese, and liver. A lack of vitamin A can produce "night blindness" (inability to see in dim light) and retard normal growth. It may also cause intestinal disorders and make the eyes more susceptible to infection.

The B complex vitamins are water soluble and nearly all are found in the same foods. The most common and well known of the B vitamins are B_1 (thiamine); B_2 (riboflavin); a constituent of B_2, niacin; and B_{12}. Major sources are all meats, fish and poultry, liver, dairy products, eggs, all cheeses, bread, cereals, and white potatoes. Pork is particularly high in thiamine and liver in riboflavin and niacin. A lack of thiamine can lead to nervous disorders, such as beri-beri or polyneuritis. A deficiency in riboflavin contributes to retarded growth, muscular weakness, and a lack of vitality, while a lack of niacin may cause pellagra, a disease characterized by eruptions on the hands, face, sore tongue, disturbed digestion, and mental disorder. Anemia will usually result from a lack of the B_{12} vitamin, the only vitamin containing a metal—cobalt. *Anemia* is a condition in which the red corpuscles in the blood are reduced or deficient in hemoglobin, causing pallor, lack of energy, shortness of breath, and palpitation of the heart.

Vitamin D is necessary to help the body absorb calcium from the digestive tract and to build calcium and phosphorous into bones. Milk, butter, cod liver oil, and egg yolk are the chief sources, but it is also manufactured in the body through action of sunlight. A lack may cause rickets, a disease that softens the bones and causes various deformities.

Another vitamin that has gained commercial popularity is vitamin E which is a light fat soluble and is found in the green vegetables and the oils of wheat seed germs. In white bread, the vitamin is often removed through processing. A deficiency can lead to impotence and sterility, especially in males. This, and the fact that it can be stored, is perhaps the reason for its recent popularity as a dietary supplement.

Vitamin K is another fat soluble found in spinach, alfalfa, and the green leaves of most vegetables. Its use is as an aid to the process of blood clotting in open wounds. Where there is a deficiency, a hemorrhage may occur from even a small wound.

Water. Though water carries no food value and strictly speaking is not a nutrient, it is important as the medium in which many of the chemical changes in the body take place. Water makes up 60 to 65 percent of an adult's body weight and is necessary in the amount of two to three quarts a day. Some of the principal functions of water are to assist in the digestion of foods, elimination of waste, glandular secretion, and in forming blood plasma. Water is lost from the system through breathing, sweating, and urine excretion, and must be constantly replaced.

proportion of nutrients required

It is difficult to give precise figures on the percentage of protein, carbohydrate, and fat that should be included in a diet. In general, carbohydrates should account for from 46 to 52 percent of the total calorie intake; fats from 20 to 35 percent. and protein from 12 to 40 percent, with a safe minimum of 0.45 grams per pound of body weight recommended for normal adults by the Food and Nutrition Board of the National Research Council. Protein proportions should be in the higher range for pregnant and lactating women, young people still in the growing stages, and those in exercise programs aimed at increasing lean muscle mass with a proportional decrease in fat. A diet selected from the variety of foods available in the four basic food groups will normally supply all the vitamins and minerals a person needs.

the four basic food groups

The essential foods necessary to supply all the nutrients for the repair, growth, and energy needs of the body are classified into four major groups. Food from each group should be included in both a normal and weight control diet. The foods listed under each group provide a wide variety from which to obtain the necessary carbohydrates, proteins, fats, minerals, and vitamins discussed.

Meat group. Included in this group are meat, fish, cheese, beans, dry peas, eggs, nuts, and poultry. These foods are all high in protein. Iron, thiamine, riboflavin, and niacin are also provided from these sources. Daily intake should include two or more servings, preferably with each meal supplying some protein from these sources. Notice that the protein does not have to come from meat only.

Vegetable-fruit group. Servings from this group should amount to three or four a day, with both green and yellow vegetables included. Although fruits and vegetables can be interchanged, it is best to include both. Fruits provide an excellent source of simple carbohydrate energy, vitamin A (yellow fruits), and vitamin C (citrus fruits, tomatoes, strawberries, canteloupe). The vegetables are a primary source of vitamins and minerals, particularly vitamins A and C. Potatoes and corn provide starches for carbohydrate energy but involve a much more complicated digestive process than fruit carbohydrates. They include other nutrients more resembling those in the bread-cereal group.

Bread-cereal group. This group includes enriched or whole grain breads, cooked or dry cereals, cornmeal, crackers, flour, grits, macaroni, spaghetti, rice, rolled oats, potatoes, corn, and beans. Three or four servings including bread or cereal should be included each day. Though the primary contribution of this group is carbohydrate energy in the form of sugar and starches, it also contains protein, vitamins, and minerals, including the B vitamins, vitamin K, and iron.

Milk group. Though children have the greatest need for this group, adults also have a continual need for dairy products. Whole milk is most conveniently used, but other forms can supply the same values, such as buttermilk, powdered milk, cottage cheese, ice cream, and cheddar cheeses. The group is a major source of protein, fat, vitamin D, vitamin B (especially riboflavin), and calcium.

Fats. In addition to the four major food groups, it is necessary to add another that is usually listed separately in a weight control diet in order to insure adequate nutrition—fat. It is also the subject of much controversy. Generally, a minimum of one tablespoon daily should be planned in the form of vegetable or olive oil, butter or margarine and used on salads, breads, or in cooking. The vegetable oils may come from safflower, sesame, soybean, or corn. Other sources of fat include animal fats, whole milk, and cream.

Much controversy still exists over the amount and type of fats that should be included in the diet. The current recommendation is to limit total fats of any kind to no more than 35 percent of the total calories and saturated fats (animal fats, butter, lard) to less than 10 percent, or about one-third of the total fats.

the role of exercise
in weight control

Since there are two parts to the weight control equation—calorie intake and calorie expenditure, it should be obvious that there are three ways in which weight loss or gain can be accomplished: (1) change calorie intake; (2) change calorie expenditure; or (3) change both. Too often, when attempting to lose weight, people tend to think only of reducing calorie intake (usually by some severe, self-designed "crash diet") and to ignore the other half of the equation. The result is usually a weight loss, but under a feeling of great deprivation and resentment. When the weight loss has been achieved, they quickly revert to their former eating habits and are soon repeating the process.

Since the body can adapt to calorie restriction by becoming more metabolically efficient, this can partially negate the weight reduction capabilities of diet alone. On the other hand, including exercise would not only burn off more calories in itself, but also help to raise the body's metabolic activity at all levels of activity, thus helping to burn off even more calories. As for the argument that an increased appetite will offset the calorie expenditure value of the exercise, we have already seen that appetite does not begin to adjust proportionately to the level of activity until a person has become adapted in the "normal range of activity" (minimum of one hour daily of vigorous physical activity to the point of exhaustion). Rather, in persons accustomed to sedentary living, appetite will usually decrease. Other studies with rats showed that exercise over a period of time does involve sufficient calorie expenditure to reduce weight without causing a proportionately larger appetite increase.

The fat content in body tissues also appears to be closely related to activity patterns. For example, Mayer has demonstrated in many studies that the difference in the fat content in the tissues of obese and nonobese animals and humans was not so much a difference in calorie consumption as in energy output. These studies and the fact that inactivity usually preceded the development of obesity led Dr. Mayer to conclude that physical inactivity must be considered as the most important predisposing factor in its cause.[5]

In addition to the value of daily exercise as a direct weight control measure are other benefits to overall fitness. With regular activity, not only is excess energy burned, but improved muscle development and tone will take place. Body fat percentage will decrease. Endurance will improve and the chronic fatigue which plagues the unfit will disappear. The release of tension provided will not only be an aid to relaxation and rest, but possibly counteract any tendency towards emotional over-eating or tension smoking.

For all these reasons, the overwhelming conclusion of research in recent years is that the most effective way of taking off weight and keeping it off is through a program that combines exercise and diet and

[5] Jean Mayer, *Overweight* (Englewood Cliffs, N.J.: Prentice-Hall, Inc., 1968). See especially Chapter 5, "Activity and Weight Control, pp. 69–83.

which is designed so that a gradual weight loss is achieved through the accumulation of a relatively small daily calorie deficit. This long-range combined method, has the effect of not only ultimately bringing weight down to optimum level, but of establishing new and lasting eating and activity habits—the key to permanent weight control.

establishing the weight control program

Since weight loss is the most common concern in a weight control program, the following discussion centers on planning to bring about a safe and effective body weight and fat reduction and to maintain that reduction. For those whose object is to increase body weight, the basic technique of planning is similar, except for the reversal of procedure.

The first step in planning for a weight reduction is to determine the energy requirements for maintaining your present body weight. The next step is to determine the amount of weight loss desired and the daily calorie intake deficit required to bring about that reduction over a period of time. Finally, the plan for bringing about the deficit and the gradual reduction must be decided upon: reduction of calorie intake, increased calorie expenditure, or preferably a combination of both. All that remains is the implementation of the plan and making the necessary adjustments as progress is evaluated.

determining energy requirements

Energy requirements vary with the activities in which an individual is engaged and are influenced by body proportions and gender. The energy needs are for three major purposes: internal work or basal metabolism, external work (that above and beyond basal metabolism), and the energy cost of food intake.

Basal metabolism is a continuing action—twenty-four hours a day, with the average energy requirements varying from 1,500 to 1,800 calories a day for men and from 1,000 to 1,500 for women. One formula for determining basal metabolism is "weight × time." For example, 1 kilogram of body weight burns roughly 1 calorie per hour. Since 1 kilogram is equal to 2.2 pounds, a 154-pound man would weigh 70 kg. (154 divided by 2.2). To determine his basal metabolism energy requirements, we would simply multiply 1 calorie × 70 kg. × 24 hrs., giving us a value of 1,680 calories. (In determining the basal calorie needs for a woman of the same weight, a 10 percent reduction of this calculated value would be made: 1,680 − 168 = 1,512). To this figure would be added the total calorie cost of the normal pattern of daily activities above resting level, determined by averaging out the time spent in various activities for several typical days and calculating the energy expenditure from an appropriate calorie expenditure chart. An additional 10 percent for the energy cost of food intake would be then added to arrive at the total needs figure.

This method is fine, but very time consuming and requires a great deal of calculation. The method we prefer is a most efficient, equally accurate, and less troublesome procedure for determining daily calorie needs. More important, in combination with one of the several exercise and weight control diet plans, it can help develop a long-range plan for weight control.

the calorie calculator

Fig. 11 shows a Calorie Calculator which can be used in two ways. Following the six steps outlined on the chart, determine your current needs with respect to maintaining present body weight. Then, refigure the calorie needs required to maintain your "desired weight," using your body type and body weight and fat analysis taken during the fitness evaluation as your guide in determining what you should weigh. This recalculation will produce a lower recommended calorie intake (assuming a lower weight is desired). By adjusting your total intake to the lower figure, weight loss will occur gradually until the desired level is reached. An example is given below.

Following the steps on the chart, we find that a person weighing 160 pounds (Step 1) and measuring 5 feet 8 inches tall (Step 2) would have a body surface area of 1.87 square meters. Assuming the person to be a male and twenty years old, Steps 4 and 5 would indicate a basal metabolic need of 1,820 calories. After consulting the Activity Level Guide, which follows the Calorie Calculator, a 60 percent activity level is estimated. Carrying out Step 6, then, shows that a total of 2,900 calories is needed to maintain a current weight of 160 pounds.

On the basis of the body type and weight and fat analysis, however, let us now assume that a desirable weight of 140 pounds is decided upon. Using this new weight projection and repeating Steps 1 to 5, we find that the body surface area is reduced to 1.75 and the basal metabolism requirements to 1,700 calories. Retaining the 60 percent activity level, Step 6 now shows a total calorie need of 2,650 to maintain the 140 pound weight. By restricting calorie intake to 2,650 calories per day, a 250 calorie deficit would result between the amount needed to maintain 160 pounds and that needed to maintain 140 pounds. On the basis of 3,500 calories to 1 pound of fat, one pound would be lost approximately every fourteen days and the desired weight would be reached in 280 days.

However, as long as a balanced diet is followed, it is safe to reduce at a faster rate. Generally, 2 to 3 pounds a week is considered both safe and satisfactory. This means that we can increase the daily calorie deficit to between 1,000 and 1,500 calories, if desired. Assuming a weight control diet of 1,800 calories a day was selected, the actual deficit for our hypothetical male would be 1,100 calories a day: the 2,900 needed to maintain current weight less the 1,800 in the selected diet. A weight loss of approximately 2 pounds a week would result and the desired weight of 140 pounds (20 pound loss) would be reached in approximately ten weeks. When the desired weight was reached, the intake for stabilization would then be adjusted to 2,650 calories for our hypothetical male.

activity level guide :*
for use with calorie calculator

20 percent should be added to the basal caloric needs of individuals who are bed patients, or those whose activities are confined to sitting in a wheel chair.

30 percent should be added for ambulatory patients who are in need of more rest than are normal individuals, but who are able to engage in limited physical exercise such as is obtained by short walks.

40 percent should be added for individuals of somewhat greater activity, but whose energy output is still considerably below par. In this class may be cited such persons as housewives who engage in various social activities, but who hire others to do their housework and students who are not participating in regular physical activity, but are primarily engaged in study. Usually they are nonworking students whose physical activity is primarily limited to walking to and from classes.

50 percent should be added for individuals engaged in clerical duties, various machine operators, cooks, domestics of various sorts, chauffeurs, and others doing similar semisedentary work. Students who work at clerical jobs such as the library or as laboratory assistants are included in this percentage area. Their physical activity includes about two hours of walking or standing daily.

60 percent should be added for manual laborers, truck drivers, farmers of various types, roofers, and the like. Teenage children who are overweight are included in this classification. College students who, in addition to their academic studies, participate in limited physical exercises

---

* Activity Level Guide adapted by permission of The Pacific Press Publishing Association, Mountain View, California, from *Reduce and Be Happy*, by Donald W. Hewitt, M.D., copyright Pacific Press Publishing Association, 1955.

Calorie Calculator

MALE AGES

LINE 7

FEMALE AGES

HEIGHT IN FEET AND INCHES

LINE 6

PER CENT ABOVE FASTING AND RESTING OR
(BASAL METABOLISM)

LINE 5

100 90 80 70 60 50 40 30 20 10 0

TOTAL DAILY CALORIC EXPENDITURE (TOTAL METABOLISM)

LINE 4

4000 3500 3000 2500 2000 1500 1400 1300 1200 1100 1000 900 800 700 600 500

CALORIES USED DAILY IF FASTING AND RESTING (BASAL METABOLISM)

LINE 3

3500 3000 2500 2000 1500 1400 1300 1200 1100 1000 900 800 700 600 500

SURFACE AREA IN SQUARE METERS

LINE 2

2.9 2.8 2.7 2.6 2.5 2.4 2.3 2.2 2.1 2.0 1.9 1.8 1.7 1.6 1.5 1.4 1.3 1.2 1.1 1.0 0.9 0.8 0.7 0.6

WEIGHT IN POUNDS

LINE 1

340 320 300 280 260 240 220 200 180 160 140 120 100 90 80 70 60 50 40

Step 1 Using a pin as a marker, locate your actual weight on line 1.
Step 2 Setting the edge of a ruler against the pin, swing the other end to your height on line 6.
Step 3 Remove the pin and place it at the point where the ruler crosses line 2.
Step 4 Keeping the edge of the ruler firmly against the pin on line 2, swing the right-hand edge to your sex and age on line 7, using the age of your nearest birthday for the purpose.
Step 5 Remove the pin and place it where the ruler crosses line 3. This gives you the calories used daily (in twenty-four hours) if you are resting and fasting.
Step 6 To the basal calories thus determined, add the percentage above fasting and resting for your type of acitivity, using the Activity Level Guide. Leaving the pin in line 3, swing the edge of the ruler to the right to the proper percentage on line 5. Where the ruler crosses line 4, you will find the number of calories necessary to maintain you at your present weight.

* *Calorie Calculator* reproduced and adapted by permission from the Pacific Press Publishing Association, Mountain View, California, from the book, *Reduce and Be Happy*, by Donald W. Hewitt, M.D., 1955.

Fig. 11.

such as are offered in the physical education activity courses, or intra-mural sports of a moderate nature (see list, pp. 186–87, for energy costs of various sport and recreational activities) are included, as are those stu-dents who attend dances and other social activities regularly and walk or stand about two hours each day.

70 percent should be added for individuals who engage in heavy work, such as construction work, mining, and stevedoring. In this group also are college students participating in physical education classes and indi-vidual sport and exercise activity programs on a regular, daily basis, and those on intercollegiate teams of a moderate nature with daily practices and weekly contests.

80 to 100 percent should be added for those engaged in the heaviest type of work described in the 70 percent category, and for those students participating in intercollegiate sport activities that have the highest rate of calorie expenditure as shown on the activity chart (basketball, track, etc.).

the combined diet and exercise program

In the example given, the planned weight loss was by calorie reduction only, with no change in the basic activity level. But we said earlier that a combination of reduced intake and increased expenditure was more desirable. If the young man wanted a combination method, he could plan it in one of two ways: (1) Plan to establish a higher basic activity level, according to the general guidelines given in the Activity Level Guide, and recalculate his calorie needs for the same rate of loss (allow-ing a greater calorie intake) or for a faster rate of loss (by maintaining the 1,800 calorie diet). (2) Retain the calculations made on the basis of his basic 60 percent level, refer to the Calorie Costs of Various Activities (p. 186), and choose to include in his daily program an appro-priate activity that would increase his daily calorie expenditure beyond his normal basic activity level for the estimated ten week period.

Assuming he chooses cycling at 10 mph. for an hour a day, on the basis of his current weight an additional 480 calorie expenditure would result. This would offer several advantages. First, an additional 480 calories could be added to the 1,800 reducing diet, bringing it to 2,280—certainly not a very severe diet, but one that would still result in the desired weight loss; or he could retain the 1,800 calorie diet and increase his rate of loss to approximately 3 pounds a week, shortening his weight loss program to seven weeks.

The second advantage is that by establishing and continuing such a pattern of activity, he would definitely place himself within the "normal range of activity" described by Mayer. When the desired weight loss was achieved and the reducing diet discontinued, weight stabilization could then be accomplished without bothering to "count calories," since appetite and food intake would automatically adjust to varying levels of activity. As pointed out, the continued regular exercise would also help to stabilize a high or low metabolic rate. In addition, it would provide all the other benefits of total fitness: cardiovascular endurance, muscle tone, relaxation, etc. Should the cycling program be discontinued and not replaced by an equivalent activity, it would be necessary to restrict calorie intake to no more than 2,650 daily calories in order to maintain the 140 pound weight. In combining an exercise and diet program, of course, many other activities can be used, depending upon interest, skill level, and convenience. The values given can be used to calculate the calories expenditures for any increment of time desired.

calorie cost of various sport and recreational activities

Because of the many variables in determining energy expenditure, the values listed must be considered as approximations only. They were arrived at by averaging values from a number of sources. It is well established, however, that calorie expenditure for various activities increases in direct proportion to body weight. For ease of calculation, the values are given in terms of calories per minute per pound of body weight, with an example of the hourly cost for a person weighing 150 pounds. For a young man weighing 180 pounds and running a mile in six minutes (10 mph) the calorie expenditure would be: .1 × 6 min. × 180 lbs. = 108 calories. If he were to run the mile in ten minutes (6 mph), however, the cost would be approximately 142 calories (.079 × 10 min. × 180 lbs.) Though the rate of calorie expenditure is greater at the faster speed, the total calorie burn-off is larger for the ten minute mile because of the longer duration. For purposes of weight loss, then, it would be more advantageous to extend duration rather than increase intensity.

the weight control diet

Table 8 shows some sample nutritionally sound weight control diets ranging from 1,200 to 2,200 calories a day that can be selected according to your personal needs and plan of approach. For those interested in gaining weight, various combinations of the diet plans can be used to increase the total calorie intake to the desired amount. You will notice that no specific menus are described in the diet plans (though a sample menu for a 1,200 calorie diet is shown in Table 9). Only the general food groups and amounts to be taken are given for each meal, with a

Activity	Cal/min/lb	Cal/hr/150 lb
Archery	.034	305
Badminton:		
Moderate	.039	350
Vigorous	.065	585
Basketball:		
Moderate	.047	420
Vigorous	.066	· 595
Baseball:		
Infield-outfield	.031	280
Pitching	.039	350
Bicycling:		
Slow (5 mph)	.025	225
Moderate (10 mph)	.05	450
Fast (13 mph)	.072	650
Bowling	.028	255
Calisthenics:		
General	.045	405
5BX (Ch. 3–4)	.098	880
Canoeing:		
2.5 mph	.023	210
4.0 mph	.047	420
Dancing:		
Slow	.029	260
Moderate	.045	405
Fast	.064	575
Fencing:		
Moderate	.033	300
Vigorous	.057	515
Fishing	.016	145
Football (tag)	.04	360
Gardening	.024	220
Gardening-Weeding	.039	260
Golf	.029	260
Gymnastics:		
Light	.022	200
Heavy	.056	505
Handball	.063	570
Hiking	.042	375
Hill Climbing	.06	540
Hoeing, Raking,		
Planting	.031	280
Horseback Riding:		
Walk	.019	175
Trot	.046	415
Gallop	.067	600

Activity	Cal/min/lb	Cal/hr/150 lb
Jogging:		
4.5 mph (13:30 mile)	.063	565
Judo, Karate	.087	785
Motor Boating	.016	145
Mountain Climbing	.086	775
Rowing:		
(Rec 2.5 mph)	.036	325
Vigorous	.118	1060
Running:		
6 mph (10 min/mile)	.079	710
10 mph (6 min/mile)	.1	900
12 mph (5 min/mile)	.13	1170
Sailing	.02	180
Skating:		
Moderate (Rec)	.036	325
Vigorous	.064	575
Skiing (Snow):		
Downhill	.059	530
Level (5 mph)	.078	700
Soccer	.063	570
Squash	.07	630
Stationary Run:		
70–80 cts/min	.078	705
Swimming (crawl):		
20 yds/min	.032	290
45 yds/min	.058	520
50 yds/min	.071	640
Table Tennis:		
Moderate	.026	235
Vigorous	.04	360
Tennis:		
Moderate	.046	415
Vigorous	.06	540
Volleyball:		
Moderate	.036	325
Vigorous	.065	585
Walking:		
2.0 mph	.022	200
3.0 mph	.03	270
4.0 mph	.039	350
5.0 mph	.064	575
Water Skiing	.053	480
Wrestling	.091	820

table 8
weight control diets

Meal and Exchange	List No.	Cal. Per Exchange	Protein	Carbo-hydrates	Fat	1200 Amt.	1200 Cal.	1500 Amt.	1500 Cal.	1800 Amt.	1800 Cal.	2200 Amt.	2200 Cal.
Breakfast													
Fruit	3	45	–	11 gm	–	1	45	1	45	1	45	2	90
Bread	4	70	2 gm	15 gm	–	1	70	2	140	2	140	2	140
Meat	5	75	9 gm	–	4 gm	1	75	1	75	2	150	2	150
Fat	6	35	–	–	4 gm	1	35	1	35	2	70	2	70
Milk	1	160	9 gm	12 gm	9 gm	1	160	1	160	1	160	1	160
Free Food	7	–	–	–	–	any	–	any	–	any	–	any	–
Totals		385	20 gm	38 gm	17 gm		385		455		565		610
Lunch													
Meat	5	75	9 gm	–	4 gm	1	75	2	150	2	150	3	225
Free Vegetable	2A	–	–	–	–	any	–	any	–	any	–	any	–
Vegetable	2B	30	2 gm	6 gm	–	–	–	–	–	1	30	1	30
Bread	4	70	2 gm	15 gm	–	2	140	2	140	2	140	2	140
Fat	6	35	–	–	4 gm	1	35	1	35	1	35	1	35
Fruit	3	45	–	11 gm	–	1	45	1	45	1	45	2	90
Milk	1	160	9 gm	12 gm	9 gm	–	–	1	160	1	160	1	160
Free Food	7	–	–	–	–	any		any		any		any	
Totals		415	22 gm	44 gm	17 gm		295		530		530		680

Dinner

Food	Cal	Pro (gm)	Cho (gm)	Fat (gm)	No.	Cal	No.	Cal	No.	Cal	No.	Cal
Meat — 5	75	9 gm	–	4 gm	3	225	3	225	4	300	6	450
Free Vegetable — 2A	–	–	–	–	any	–	any	–	any	–	any	–
Vegetable — 2B	30	2 gm	6 gm	–	1	30	1	30	1	30	1	30
Bread — 4	70	2 gm	15 gm	–	1	70	1	70	2	140	2	140
Fat — 6	35	–	–	4 gm	1	35	1	35	1	35	1	35
Milk — 1	160	9 gm	12 gm	9 gm	1	160	1	160	1	160	1	160
Fruit — 3	45	–	11 gm	–	–	–	–	–	1	45	2	90
Free Food — 7	–	–	–	–	any	–	any	–	any	–	any	–
Totals	415	22 gm	44 gm	17 gm		520		520		710		905
Total Calories for the day						1200		1505		1805		2195
% Protein, Carbohydrate, Fat					31, 48, 21		32, 47, 21		32, 47, 21		33, 47, 20	

Flexibility Rules: Exchanges may be transferred, combined, or substituted within these limits: (1) At least three separate meals must be eaten; (2) at least one meat, fat, fruit or vegetable serving must be included in each meal; (3) milk and bread exchanges must be distributed over at least two meals; (4) a minimum of 25% of the total daily calories must be eaten at breakfast and a minimum of 20% at lunch; (5) if desired, ½ bread exchange may be substituted by 1 vegetable or fruit serving or 1 full exchange by 2 vegetable or fruit servings or a combination (limit 1 full bread replacement). A fruit and vegetable serving may be interexchanged once, provided vitamin A and C requirements are met.

reference made to Exchange Lists which follow the table on reducing diets from which the specific foods can be selected.

All the foods in each Exchange List contain approximately the same number of calories and nutrients when taken in the amounts indicated. The great variety of foods offered through this exchange plan provides for an interesting and varied diet to meet personal taste while still assuring proper nutrition and the desired results. Some foods, such as those listed under Exchange List 2A, contain such a negligable amount of calories that no restriction is made on the amounts that can be eaten raw. For those that normally require cooking, up to 8 ounces can be taken without any adjustment in the total diet. These foods thus allow for "nibbling" or for making a satisfying "full meal" without added calories.

You will notice that a good and substantial breakfast is included in each diet plan, with a minimum of three meals planned. The digestive system operates most efficiently when it is patterned into a repetitive cycle. Furthermore, research with animals and humans shows that those who consume all or most of their daily food intake in just one or two meals have a greater tendency toward obesity; the one meal a day pattern in the evening is the worst practice. If it is remembered that for most persons, more calories are needed after breakfast and lunch than after dinner, these findings should not be surprising. Spacing meals throughout the day will also provide immediate energy and avoid late morning or afternoon fatigue and irritability caused by low blood sugar levels. If desired, you may distribute some of the allotted items throughout the day to allow for mid-morning, mid-afternoon, or evening snacks, but do not combine or skip any meals.

For those whose weight control plan calls for totals between those listed, additional food items can be selected from the Exchange Lists. Remember that the values given are approximations only, and individual responses to diets may vary. The same is true in calculating total energy requirements and calorie expenditures for various exercise activities. Adjustments, therefore, may be necessary as you progress. For example, if your weight loss under a particular diet or combination diet and exercise plan is slower than expected, you may have to drop to the next lower diet plan or increase activity. If you lose at a faster rate than 2–3 pounds a week, it would be best to change to a higher diet plan. Despite the necessary approximations in calculating calorie intake and expenditure, the margin of error is relatively small. The important thing is to develop a sound plan, based on the best of accepted principles of diet and exercise. As the plan becomes a lifelong style, the small necessary adjustments can easily be made. A summary form for planning a personal weight control plan is included at the end of the chapter, together with a short list of high calorie foods.

Each food and serving portion shown within a list contains about the same number of calories and nutrients as any other in the list. Some are slightly higher or lower than the amounts shown at the top of each list, but the differences are neglibible. When more than one serving is indicated on your plan, you may simply multiply the portions of one item by the appropriate number or you may use a combination of food servings up to the indicated amount.

In measuring foods, you will need a standard 8 ounce measuring cup, a teaspoon, and a tablespoon. Keep the measures level. Weights and volumes listed on food containers and packages will also be helpful. Unless indicated or obvious, food measures refer to cooked foods. Meats will weigh approximately 40 to 60 percent less after being cooked, depending on type and cut. For example, a half-pound of lean sirloin steak will weigh only about 3 ounces after it is cooked and be equivalent to 3 meat exchanges as listed in the diets. A small diet scale for weighing meats may be purchased at a drug store for approximately $1.00 and be used most conveniently. Unless indicated, meats and poultry should be weighed boneless.

Certain foods have been omitted from the exchange lists because they have a lot of sugar and "empty calories" (nutrients not in proportion to the high number of calories contained), and would not easily fit into a reducing diet. Candy, honey, jam, jelly, syrup, pie, cake, cookies, and sugared soft drinks are examples of foods that must be avoided. Wine, beer, and alcoholic beverages are others not usually allowed. If taken, they should be limited to a small portion. Once you have achieved your desired goal and established a sound eating and regular activity program these foods may be gradually brought in, but in moderation.

list 1 :   milk exchanges :

Calories: 160; Protein: 9 gm; Carbohydrates: 12 gm; Fat: 9 gm.

Milk, whole, 1 cup	*Milk, buttermilk, 1 cup
*Milk, skim, 1 cup (liquid)	Ice Cream, regular (10% fat), 3 fl. oz. cup**
	Yogurt, whole milk, 1 cup

6 Format adapted from *Reduce and Be Happy* by Donald Hewitt, M.D., by permission of the publisher, The Pacific Press Publishing Association, Mountain View, California. Values updated from original listings from *Nutritive Value of Foods,* Home and Garden Bulletin No. 72 (Washington, D.C.: U.S. Government Printing Office, 1970).

*Add 2 fat exchanges from List 6 when skim milk or buttermilk is used; these do not contain the same food value as whole milk. The added fat exchanges will keep the diet within the proportion of protein, carbohydrates, and fat set for adequate nutrition.

**Add 1 meat exchange A. for protein-fat balance.

list 2A: vegetable exchanges:

Negligible calories. Except for limits indicated, eat as much as you want raw at any time during the day.

*Carrot, raw (limit 1)	Green Pepper (limit 1)
Celery (limit 4 stalks)	Pickles, dill-sour (limit 1 medium)
Chard	Radishes
*Chicory	Water Cress
Cucumbers	

*High in Vitamin A

list 2B: vegetable exchanges:

Fresh, canned, or frozen; cooked and drained: Calories: 30; Protein: 2 gm; Carbohydrates: 6 gm.

Asparagus, 1 cup	Mustard Greens, 1 cup**
Beets, 1/2 cup	Onions, 1/2 cup
Broccoli, 1 cup	Peas, 1/3 cup
Brussel Sprouts, 1/2 cup	Pepper, green, 1 cup**
Cabbage, 1 cup	Sauerkraut, 2/3 cup
Cauliflower, 1 cup	Spinach, 2/3 cup
*Carrots, 2/3 cup	String Beans, 1 cup
*Escarole, 1 cup	Summer Squash, 1 cup
*Beet Greens, 1 cup	Tomatoes, 1/2 cup
*Collards, 1/2 cup**	Tomato, raw, 1**
*Dandelion, 1/2 cup	Turnips, 1 cup
*Kale, 1 cup**	Wax Beans, 1 cup

* High in vitamin A. Include at least one serving a day, unless taken under 2A.

** High in vitamin C. May be taken in lieu of recommended daily citrus fruit. If not, be sure to include it in your fruit exchange.

list 3: fruit exchanges:

Fresh, canned, cooked, dried, or frozen, so long as no sugar or syrup has been added. Calories: 45; Carbohydrates: 11 gm; some protein.

Apple, 1 small
Applesauce, 1/2 cup
°Apricots, 2 raw (about 12 per
  lb.)
Banana, 1/2 medium
Blueberries, blackberries, 1/2 cup
°Canteloupe, medium, about 5"
  diam., 1/3
°°Grapefruit juice, 3-1/2 oz. glass
Grapefruit, white, medium, 1/2
Grapes, slip skin type, 2/3 cup
°°Orange, 1 small, 2" diam.
°°Orange juice, 3-1/2 oz. glass

°Papaya, 1/2" cubes, 2/3 cup°°
°Peach, 1 medium, 2" diam.
Pineapple, diced, 1/2 cup
Raisins, 1/2 oz. pkg.
Raspberries, 2/3 cup
°°Strawberries, raw, capped, 3/4
  cup
Tangerine, 1 medium,
  2-1/2" diam.
°°Tomato juice, 1 cup

° High in vitamin A. May be used in lieu of vegetable to meet minimum requirement.

°° High in vitamin C. Include at least one serving a day unless obtained from vegetable.

list 4 : bread exchanges :

Calories: 70; Carbohydrates: 15 gm; Protein: 2 gm.

Bread, any type, 1 slice
Frankfurter, hamburger roll, 1
  (count 2 exchanges)
Cereals, cooked:
  Farina, 2/3 cup
  Oatmeal, 1/2 cup
  Grits, 1/2 cup
Cereals, dry:
  Flake type, 2/3 cup
  Puff type, 1 cup
Crackers:
  Graham, 2-1/2" square, 3

Oyster, 1/2 cup
Saltine, 5
Round, thin type, 6
Flour, all purpose, 2-1/2 tbsp.
Beans, dry, cooked, drained:
  Lima, 1/4 cup
  Great Northern, Navy, Cowpea,
  1/4 cup
Corn, sweet, 1 ear, 5" × 1-3/4"
Parsnips, 2/3 cup
Potato, white, 1 small
Potato, sweet, 1/4 cup

list 5 : meat exchanges :

Meats should be broiled, baked, or roasted and fish baked or broiled. If fried, use only fat exchanges allowed in plan which can also be used for broiling or baking. Calories: 75.
A. *Beef, pork, lamb, veal, chicken, liver, eggs, and certain fish.* Protein: 9 gm; Fat: 4 gm.

Beef, braised or pot roasted,
lean only, 1-1/4 oz.
Hamburger, lean, 1-1/4 oz.
Rib roast, lean only, 1 oz.

Steak, lean, 1-1/4 oz.
Pork roast, lean, 1 oz.
Veal, cutlet, medium fat, 1-1/4 oz.

Lamb, chop, lean only, 1-1/2 oz.
Lamb, roast, lean only, 1-1/2 oz.
Liver, 1 oz.
Chicken, flesh only, broiled,
2 oz.
Egg, 1 large°
Mackerel, Fresh Salmon, 1 oz.
Tuna, canned in oil, drained,
1-1/4 oz.

B. *Other fish, shell fish.*°°

Bluefish, haddock, shad, 1 oz.
Clams, meat only, 2 oz.
Clams, solids & liquids, 3 oz.

Crabmeat, canned, 1-1/2 oz.
Oysters, medium, 1/3 cup
Salmon, canned, 1 oz.

C. *Luncheon Meats, Cheese*

Cheese, cheddar, swiss, American,
2/3 oz.°°°
Cottage cheese, uncreamed, 2 oz.°°
Bologna, 2 slices°°°
Deviled Ham, canned,
1-1/4 tbsp.°°°

Pork link (16 links/lb.), 1 link°°°
Salami, cooked, 2 slices, 3″ diam.,
1/8″°°°
Frankfurter, 1 (count 2
exchanges)°°°

° This group is higher in saturated fats and cholesterol, particularly eggs. Do not include more than 3–4 days during the week.

°° Low in fat; add 1 fat exchange when used.

°°° Approximately 50 percent less protein and more fat for same number of calories. To keep within the recommended limit of saturated fats, restrict to lunches.

list 6 : fat exchanges :

Calories: 35; Fat: 4 gm.

Bacon, crisp, 1 slice°
Butter° or margarine, 1 tsp.
Cream, half & half, 1-1/2 tbsp.°
Cream, light, 1 tbsp.°
Cream, sour, 1-1/3 tbsp.°
Salad & cooking oils, all types,
1 tsp.°°
Lard, 1 tsp.°

Salad Dressings:
Mayonnaise type, 1 tsp.
French, 1 tsp.
Blue cheese, 1 tsp.
Thousand Island, 2 tsp.
Mayonnaise, 1 tsp.
Dietary, low-cal, any amount (Do
not count as an exchange.)

Olives, green, 8 med. or 4 giant°°°
Olives, ripe, 6 small or 4 large°°°

° Higher in saturated fats. Do not use more than 3–4 days a week.
°° Low saturated oils include: corn, olive, safflower, and soybean.
°°° Negligible saturated fats.

list 7 : free food exchanges :

Negligible in Calories.

Bouillon, clear	Mint
Coffee, tea	Mustard
Flavorings, unsweetened	Nutmeg
Gelatin, plain°	Onion seasoning
Cranberries	Pepper and other spices
Chopped parsley	Salt
Garlic	Seasoned salts
Lemon for seasoning	Vinegar

° A variety of tasty gelatin desserts can be made by combining it with lemon, coffee, or various fruit exchanges.

table 9
sample menu: 1200 calorie diet

	List	Amt.	Cal.	Actual Menu
Breakfast				
Fruit Exchange	3	1	45	1/2 grapefruit
Bread Exchange	4	1	70	1 cup skim milk
Meat Exchange	5	1	75	1 poached egg on buttered
Fat Exchange	6	1	35	toast, with 2 slices of crisp
Milk Exchange	1	1	160	bacon
Free Food Exchange	7	any	–	coffee with skim milk from
Total Calories			385	milk exchange
Lunch				
Milk Exchange	1	–	–	cooked salami and cheese
Free Vegetable Exch.	2A	any	–	sandwich with mayonnaise
Vegetable Exchange	2B	–	–	and lettuce.
Fruit Exchange	3	1	45	a sliced raw carrot and 1 cup
Bread Exchange	4	2	140	of fruit gelatin mixed with
Meat Exchange	5	1	75	1/2 cup of diced pineapple.
Fat Exchange	6	1	35	
Free Food Exchange	7	any	–	
Total Calories			295	

table 9 (cont.)

	List	Amt.	Cal.	Actual Menu
Dinner				
Milk Exchange	1	1	160	small salad with low calorie
Free Vegetable Exch.	2A	any	–	dressing
Vegetable Exchange	2B	1	30	3-3/4 oz. of beef potroast,
Fruit Exchange	3	–	–	lean only (approximately
Bread Exchange	4	1	70	1/2 portion of precooked
Meat Exchange	5	3	225	1 lb. cut)
Fat Exchange	6	1	35	1 cup asparagus
Free Food Exchange	7	any	–	1 small baked potato with
Total Calories			520	1-1/3 tbsp. sour cream and

Total for Day: 1195

chives.
3 fl. oz. cup of vanilla ice
cream
coffee, black

Evening Snack
1 cup bouillon and 1 cup plain gelatin mixed with lemon juice.

summary of basic rules for sound weight control

1. Eat balanced meals, structured from the four basic food groups. All the nutrients needed can be found in the daily foods we eat, if selected from the correct sources.

2. Don't skip meals. Eat regularly at about the same time each day. Three to five meals daily is a good practice to follow, depending on activity schedule, but the actual number is not as important as the regularity of eating times.

3. Eat at least one-fourth of your daily calorie intake at breakfast. Blood sugar is lowest in the morning and the body needs greatest. Breakfast should include protein.

4. Eat slowly and enjoy each meal. One of the major dietary faults of Americans is rushing through meals. Failure to properly chew food makes digestion more difficult, can lead to dental problems, and can cause overeating.

5. Save the juices, liquids, and fats left over from cooking. Many vitamins rest in this base, along with some minerals. Use the juices as dressings over foods.

6. Keep total fats under 35 percent of the total calorie intake and saturated fats under 10 percent, or to one-third of the total fat intake.

7. Avoid fat diets and remember *Calories Do Count*—both calorie intake from food and calorie expenditure from activity.

8. The key to permanent weight control is projecting weight change on a long-term basis and developing sound, life-long habits of diet and exercise.

personal weight control plan

Date _____ Age _____ Ht. _____ Wt. _____

Basic Activity Level Percentage: _____ %

Calories Needed to Maintain Present Wt.: _____

Desired Wt. for Body Type: _____ Desired Wt. Loss or Gain: _____

(indicate by + or − number of pounds)

Daily Calorie Intake Needed to Lose or Gain 2 to 3 lbs. per Week:

A. Calories Needed to Maintain Present Weight: _____

Calorie Deficit or Addition: _____ | 1,000 to 1,500

(indicate − or +)

Net or Total Calorie Intake Needed to
Lose or Gain 2 to 3 lbs. Per Week: _____

B. Selected Diet Plan or Combination (Calories): _____

Summary of Exchanges and Notes on Diet Plan

Exchanges	*List*	*Amt.*	*Meal Plans Should Include:*
Milk	1		1. At least one meat, fat, fruit or vegetable exchange in every meal.
Vegetable	2A	_____	
Vegetable	2B	_____	2. At least one milk and one bread exchange in two meals.
Fruit	3	_____	
Bread	4	_____	3. At least 25% of total daily allowance at breakfast.
Meat	5	_____	
Fat	6	_____	4. At least one citrus fruit exchange,
(Avoid High Calorie Foods)			one green and one yellow vegetable each day.

C. Actual Daily Calorie Deficit or Addition
(A minus B, or B minus A): _____

D. Approximate Weekly Loss or Gain $(C \times .002)$** _____ lbs.***

E. Approximate Weeks to Reach Desired Weight
(Total Loss ÷ Wkly. Loss): _____

---

* 1,200 minimum, even if rate of loss is slower.

** (Daily Deficit) × (Days ÷ 3,500 Cal. = Lbs. Loss; 7 days ÷ 3,500 Cal. = .002.

*** For a faster rate of loss, add an additional activity from list on pp. 186–187, above those regular activities considered in determining your Basic Activity Level Percentage. You may also use an additional activity to increase your food intake by the equivalent amount expended, and still maintain the same rate of loss. Also, though the actual deficit will decrease slightly as weight loss occurs (lower basal metabolism requirements), the difference is not significant enough to consider (approximately 10 cal/day per lb. loss; 70 cal/day after a 10 lb. loss, etc.).

some high calorie foods

Beverages

Beer, 12 oz.	150
Carbonated water, 12 oz.	115
Cola type, 12 oz.	145
Gin, Rum, Vodka, Whiskey, 80 Proof, 1-1/2 oz.	100
Manhattan, Whiskey Sour, 2 oz. reg. cocktail	140
Martini, 2 oz.	160
Wine, dry, 12%, 3-1/2 oz.	85
Wine, sweet, 22%, 3-1/2 oz.	140

Tidbits

Peanuts, roasted, 1 cup	840
Potato Chips, 10 med.	115
Pretzels, stick, reg., 5	10
Popcorn, with oil and salt, 1 cup	40
Popcorn, sugar coated, 1 cup	135
Pizza, 1/8 of 14" pie, cheese	185

Macaroni, Noodles, Spaghetti

Macaroni, cooked:	
Firm, 1 cup	190
Until tender, 1 cup	155
Macaroni and Cheese, baked, 1 cup	430
Noodles, egg, cooked, 1 cup	200
Spaghetti, cooked:	
Tender stage, 1 cup	155
In tomato sauce with cheese, cup	260
With meatballs and tomato sauce, cup	330

Sweets and Desserts

Candy:	
Caramels, plain or chocolate, 1 oz.	115
Chocolate, milk, plain, 1 oz.	145
Chocolate-coated peanuts, 1 oz.	160

Chocolate bar, ave., 2 oz.	240–275
Jams and Preserves, 1 tbsp.	55
Sirups, table blends, 1 tbsp.	60
Sugars:	
White, granulated, 1 tbsp.	40
Cakes:	
Cupcake, small, with icing	130
White, 2 layer, with icing, piece, 1/16 of 9" diam.	250
Cookies:	
Brownies, with nuts, 1	95
Chocolate Chip, 1	50
Fig bar or sandwich type	50
Crackers:	
Graham, 2-1/2" square, 4	110
Danish pastry, plain, approx. 4-1/4" diam., 1	275
Doughnuts, cake type, 1	125
Pie:	
Apple, cherry, 1/7 of 9" diam., 1 piece	350
Custard, 1 piece	285
Lemon Meringue, 1 piece	305
Pecan, 1 piece	490
Toppings:	
Cream, whipped, 1 cup.	400
Pressurized, 1 cup.	155

Other

Creamers, powdered, 1 tsp.	10
Cream, liquid, heavy, 1 tbsp.	55
Cream, liquid, light, 1 tbsp.	30
Custard, baked, 1 cup	305
Ice Cream, reg, 10% fat, 1 cup	255
Pudding, chocolate, 1 cup	385
Pudding, vanilla, 1 cup	285
Sherbet, 1 cup	260

8

handling
stress and tension

Because of the increased amount of stress and tension associated with living today, and the damage caused by it, the ability to relax has become an important part of fitness. *Stress* refers to the amount of strain or pressure exerted on a person. It implies a demand placed upon the body. *Tension* refers to the nervous anxiety, usually accompanied by muscle tightness, that results from certain kinds of mental and emotional stress.

A certain amount of stress is a normal, necessary part of everyone's life and can't be avoided without failing to live fully. It is only when a particular stress is applied too long that difficulties arise. Knowing how to avoid these harmful stresses or if unavoidable how to relieve the tension created by them is the primary subject of this chapter.

nature and effects of stress

As noted, stress or pressure implies a demand placed on the body. The demand may be physical, mental, or emotional, and of varying types and intensities. Physically, the stress may be as small as rising out of bed in the morning or as intense as a hard game of tennis or a full day of vigorous physical work. Fatigue, illness, or accident are other forms of physical stress on the body. Mentally, the stress may be reading a book, solving a complicated problem, or worry over an unsolved problem. Emotional stress may include the joy of seeing a long-separated loved one, the anguish of seeing a loved one die, the excitement of watching good drama or a football game, or feelings of anger, hatred, fear, or frustration aroused by a variety of situations.

body response to stress

The body responds—or attempts to respond (barring disease or injury to vital organs)—to any type of stress in the same manner. Interpreting the stress as a call to action, it mobilizes itself through the interaction of the nervous and glandular systems which control the level of body activity. Adrenalins are secreted, stored energy is released, blood sugar level rises, and heart rate and blood pressure increase. When the stress is removed, calming hormones are released and the body returns to normal.

stress as a necessary energizing force

Obviously, a certain amount of stress is a necessity of life. It is an energizing force without which we would be only half alive. As an illustration, consider how you feel when you wake up in the morning after a good night's sleep. The body is just awakening from its least stressful state. You're groggy and uncoordinated. As you are gradually subjected to the stress (pressure) of having to respond to the stimuli of light and sound, having to rise, move about, and prepare for the day's activities, the energy-producing forces are released and you "come alive" —some of you more slowly than others. Similarly, the stress of an enjoyable interpersonal encounter, job, or play activity—regardless of difficulties—is essential for a full and satisfying life.

effects of stress

According to Hans Selye, a leading authority on stress at Montreal's Institute of Experimental Medicine and Surgery, it is the continual adaptation of the body to various stresses that ultimately exhausts the body's ability to adapt, wears out vital organs, and leads to a variety of illnesses until death follows.

Keeping the body and vital organs in good condition through proper medical care, exercise, diet, and sound health habits can help the body withstand longer the various stresses it is necessarily subjected to during life. But too much stress—of any type—speeds up wear and tear and leads to an earlier pattern of illness and death. It is this prolonged stress that leads to difficulties and which must be avoided.

Overstress. Exhaustion from physical or mental overwork is an obvious form of overstress for most people. Not so obvious, however, but of even greater danger are other types of prolonged mental and emotional stress. The most harmful of these are worry, anxiety (an uneasy feeling of apprehension), fear, anger, hatred, and frustration brought about by a variety of situations.

Psychiatrists say that at the root of these types of mental and emotional stresses is the feeling of being "threatened," whether real or imagined. Some involve a threat to the basic need for security (illness, war, natural disaster, loss of job, lack of education, etc.). Others threaten basic or socially created needs or desires, such as the desire for acceptance, belonging, recognition, prestige, social position, wealth, etc. Since the body responds to messages received from the brain, but is unable to distinguish between the stimuli which triggered the message, its reaction is the same whether the threat is real or imagined. The "fight or flight" response is elicited and it mobilizes itself for action (increased adrenalin flow, energy release, heart rate, etc.). Even if the energy buildup is released by physical action, a toll is taken in the form of an unnecessary expenditure of energy and increased wear on body organs. If unreleased, the result is a tension buildup which takes an even greater toll.

tension buildup and effects

The increased amount of adrenalin in the blood tends to tighten and restrict both the smooth and striated muscles of the body, including the heart, arteries, and small blood vessels. In addition, it increases the viscosity of the blood and shortens its clotting time—undoubtedly a protective mechanism in case of injury in the "fight." The effect is not only to increase the work of the heart, but to create the greater possibility of a blood clot forming anywhere in the vascular system. Arteries not only tighten, but are further narrowed by deposits of fat drawn from body stores for energy which goes unused. The likelihood of developing atherosclerosis leading to hypertension and heart disease is obviously increased. The constriction of the smaller arteries is often responsible for headaches, fatigue (from a decreased oxygen supply), and skin eruptions.

The tightening and squeezing down hard of the stomach muscles is responsible for over half the complaints of "ulcer pain." Similar muscle tension pains are common in the intestines, gall bladder, and in the large muscles of the neck, chest, upper and lower back, and extremities.

Such ailments as colitis, gastric ulcers, and chronic diarrhea are also known to be caused by the release of stomach acids and other body chemistry changes resulting from severe emotional stress.

Mental and emotional breakdowns are still another result of the prolonged inability of persons to handle the stress and tension in their lives and to obtain relief from them. One of the most common manifestations of this inability is the manic-depressive state in which a person alternates from "high" periods of excitement and frantic activity to "lows" of extreme depression and relative immobility. Severe or frequent moods of depression, even if not associated with a "breakdown," appear to be simply an attempt by the body to escape or withdraw from stressful situations. Other minor personality disorders, such as constant irritableness and chronic impatience, are other manifestations of stress and tension.

removing and reducing
stress and tension

No matter how well intentioned, the most fruitless advice one can give to a person who is overworking, worried, or upset by any emotion is to say: "Don't work so hard," "Stop worrying," or "Don't be upset." Obviously, if it were that easy, there would be no problem.

What would be more helpful is to convince the person his actions are leading him to self-destruction and to ask: "Is it really worth it?" If convinced, the person may then learn to ask himself the same question when caught in an emotionally tense situation, and this may be the beginning of control.

More helpful still would be to help him examine himself and situations in his environment that may be the cause of the harmful overstress. Since behind every stress there is a cause, identifying and removing it would be the best way of removing the stress and tension created by it.

Where impossible to remove the cause completely, the emphasis must then shift to removing or reducing the tension through appropriate exercise, recreation, specific physical relaxation techniques, and proper rest.

removing causes

In some cases (worry, anxiety, fear), the cause may be an attitude or pattern of response that reflects deep feelings of insecurity or lack of spiritual values (faith, trust) which tend to wrongly or overinterpret various situations as "threatening" when in fact they are not. Identifying and removing such deep psychological and spiritual causes usually requires the services of professionally trained personnel. Sometimes, just talking out troubles with someone close or someone trusted and respected can be helpful.

Damaging frustration, anger, and hatred also reflect an unreal "threat to self" caused by insecurity or a failure in the spiritual area. With respect to the serious effects of anger and hatred and the role that spiritual values may play in controlling them, Selye says: "Thus, 'Love Thy Neighbor' is one of the sagest bits of medical advice ever given."[1]

common psychological causes

Other psychological factors that can cause stress and tension, though stemming from deeper causes, are more easily identifiable and therefore subject to control.

Too many wants or desires. Whether in the area of acquisition or achievement, too many wants or desires or those that are beyond the capacity of an individual to fulfill lead only to frustration and other harmful emotional stresses. Balancing wants and desires with a realistic appraisal of capacities is one way to minimize these stresses. Applying a scale of values by which to decide what is necessary and important and what is relatively unimportant is another way to help provide this balance and distribute energies accordingly.

Expecting freedom from difficulties. We have already seen that a certain amount of stress is normal and needed by the body as an energizing force. So it appears to be with difficulties too. Without them, there would be little to overcome and bring out the best that is in us. Furthermore, history and experience have shown that never has there been a time or life that did not have its problems. Therefore, expecting life to have its problems and difficulties and to involve "ups" and "downs" can help to avoid compounding them.

Expecting perfection. Closely related to balancing wants and desires with capacities and expecting difficulties is not expecting perfection from yourself or others in all situations—a truly impossible accomplishment. As humans, failure and mistakes in judgment and action must be expected as the normal consequence of the frailties and limitations of human nature. It is partly this recognition that helps a person respond to failure or mistakes with acceptance and calmness rather than with condemnation of self or others, with its attendant frustration and anger.

---

1 From "How to Avoid Harmful Stress," by J. D. Ratcliff, *Reader's Digest,* July, 1970, p. 80. Reprinted by permission from *Today's Health,* July, 1970. Copyright 1970 by *Reader's Digest.*

Certainly, we can always strive to do our best in all situations, but always with the recognition that we have "feet of clay" and that even our best will fall short of perfection.

Excessive drive and ambition. Excessive drive and ambition is often related to having too many wants or desires or sometimes to a futile attempt to overcome deep feelings of insecurity or lack of self-worth. A certain amount of drive and ambition is, of course, essential for success. But to an excess, it can result in an addiction to work that is equally as dangerous as drug or alcohol addiction. Once a "slave" to work, it is only a matter of time before the resultant and continuous stress leads to serious disorders. For such addicted individuals, allowing some time for diverting recreation can be the best kind of life insurance policy.

Negative thoughts. The body responds to messages received from the mind, but is unable to distinquish between the stimuli that triggered the message. Whether the messages are sent in response to an actual situation or triggered by mere thought, the body's reaction will be the same. For example, whether one engages in an actual argument or merely carries it on in his mind, the same harmful emotions are elicited. At the same time, positive thoughts—even in actually threatening situations—remove the threat and calm the body.

Since thoughts can be controlled, practice is possible in learning to shift from negative to positive thoughts. This is not only the basis of the art of relaxation practiced in certain forms of yoga, but of the calming effect of religious meditation. It also explains the serenity often shown even in the face of real danger by religious people of all faiths. When tempted by negative thoughts to engage in an unpleasant argument with a person, pausing to ask "Is it really worth it?" can be the first step in calming the body and controlling thoughts.

Indecision. Probably one of the most common causes of psychological stress is indecision and doubt. Indecision, by its very nature, involves conflict—weighing alternatives and an opposing pull of forces on the will. Until the will is moved either way, the conflict and stress remain. Again, the body doesn't know or care whether the decision is a major or minor one. Its response will be the same. Since decisions are basic to human nature, they cannot be avoided. But we can and must control the amount of time given to them.

The length of time given to weighing alternatives, of course, depends on the importance of the decision. Certainly, deciding on marriage, job, or college would require more deliberation than deciding on a pair of shoes. Again, having a set of values to determine the relative importance of decisions is one way to restrict the amount of time spent in deliberation. It is amazing how much time and energy some people waste in trying to decide unimportant matters that just aren't worth it.

In any event, it should be remembered that every decision necessarily involves the rejection of one alternative for another based on what an individual feels will bring him the greatest good. Whether or not the decision proves good or bad, however, cannot be known until *after* it is made, *not before*. Prior to it, only the *possible* consequences can be known and nothing is accomplishd until the decision is actually made. After giving a reasonable amount of time to a decision, based on its relative importance, and seeking advice where appropriate, one must then force himself to make up his mind, stick to it, and let the consequences run their course—right or wrong. The feeling of relief that follows will be a reward in itself, and chances are most of the decisions will be right or at least correctable in the long run. Learning to force small decisions will not only save time and energy, but it will be good practice for the larger ones.

Living in the past. Living in the past is related to rehashing decisions which appear to have been wrong. It is expressed in the saying, "If only I had done so-and-so. . ." Besides reliving the original conflict, there is the additional stress of frustration from the impossible task of trying to "turn back the clock." Methods of avoiding such frustration include accepting the consequences of decisions, having patience and allowing time for the full consequences to unravel, learning from the mistakes of past decisions, and focusing only on changes that are possible from present and future decisions.

situational causes

Blending in with these psychological factors as causes of stress and tension are certain situations which all of us find ourselves in at various times and which are even more easily controllable.

Too many unsolved problems. When too many unsolved problems accumulate, a basic rule to remember is that they cannot all be solved at the same time. Furthermore, chances are that if closely examined, not all will be of equal importance or priority. It is only when all the problems are seen cumulatively that perspective is lost and they take on the magnitude of being impossible to handle. The result is often paralysis and further accumulation of difficulties. Taking time to determine the relative importance of each, setting up priorities, and systematically taking on one at a time is the only way to gain perspective and successfully cope with them. It is surprising, too, how many times problems on the bottom of the list are solved by the solution of prior ones, or simply by the passage of time and the interaction of other events beyond our doing. In handling individual problems, talking them over with another person is often helpful in gaining perspective, insight, and suggestions on how to cope with them. Finally, when bogged down on a particular problem, diverting the mind and getting away from it for awhile is one of the best means of not only regaining perspective, but also of reenergizing the mind. It is here that exercise and recreation play key roles.

Uncontrollable problems. Many people bring on needless stress by worrying about problems over which they have little or no control. For example, while it is our human obligation to show concern and assist where we can in solving the problems of others and the world around us, unless a solution lies in our direct control, our efforts are limited. By recognizing and accepting this limitation, unnecessary frustration can be avoided. Unexpected illness, natural disaster, and events controlled by the actions or decisions of others—even though they may directly affect us—are similarly beyond our control for the most part. A recognition of this type of stress and the need to handle it is reflected in a familiar prayer attributed to St. Francis of Assisi as early as the 13th century: "Lord, grant me the serenity to accept the things I cannot change; the courage to change the things I can; and the wisdom to know the difference."

Too many or difficult responsibilities. Accepting too many responsibilities or those for which we are unprepared is sometimes necessary and unavoidable, but just as often can be of our own making. In either case, a recognition of our capacities and limitations, a weighing of what is important and unimportant, setting up priorities, and not expecting perfection are essential for minimizing undue stress and tension in these situations. In some cases, the wisest and most courageous thing to do may be simply to recognize our inability to handle certain responsibilities and change the situation.

The need for adequate rest and pleasurable mind-diverting recreational activities, even though for short periods, is even more important during times of heavy responsibilities. Planning and taking time for it is a responsibility that should be considered of equal importance to the highest and is the only way to find the time we often say is lacking. If even the President of the United States, who carries some of the greatest burdens of the world on his shoulders, can find time for a relaxing walk, a round of golf, a swim, or a movie, then certainly most of us can.

Hostile environments. People react to hostile home or job situations in a variety of ways. Some simply shrug off difficulties and disagreements with superiors, coworkers, spouses, or parents. Others repress their feelings and develop a variety of illnesses. If a situation constantly preys on the mind, however, something ought to be done. The best approach is direct communication and an airing of real or apparent difficulties or causes of concern. Quite often, the mutual understanding resulting from this approach is enough to resolve the stress. Where difficulties are impossible to resolve or where they persist, it may be better to find a new job, a competent marriage counselor, or move out of the home rather than risk serious illness.

Disorganization. Organizing and scheduling our lives—not rigidly, but so that a fairly routine pattern can be established for handling everyday responsibilities can help to avoid much of the stress that comes from unnecessary confusion, rush, and worry. We are all familiar with the tensions that can develop in a situation where we don't know what to expect. Lack of routine and organization can be looked upon in the same way. The mind and body never know what to expect. In a sense, they are always on an "alert and standby." By establishing a fairly routine pattern for handling our normal everyday tasks and breaking down our work responsibilities in an orderly fashion (by writing them down if necessary), mind and body can become conditioned to a regular pattern of demands. Once used to doing things routinely, people are often surprised at how much time they have to do the things they really want to do.

Rushing and procrastination. Related to disorganization are stress situations involving unnecessary rushing and procrastination of required but unpleasant tasks. Putting off unpleasant tasks creates a double stress: one from dreading it, the other from actually doing it. A good rule to follow here is simply to reverse the procedure: get the unpleasant jobs out of the way first and save the enjoyable ones for last.

Rushing to complete something or to get somewhere on time is another multiple stress situation, involving physical stress plus frustration and anger. Anticipating delays and planning for an earlier completion or arrival time is an effective way to avoid these stress situations.

Boredom. Another source of tension in life is boredom and the uneasy, restless feeling it brings. In today's automated society, much of the challenge and meaning of work has been lost, with the result that many people feel useless and unwanted. The same is true of the retired businessman or the woman who suddenly realizes that her children have grown up, left home, and no longer rely on her. Finding something to do that has meaning and purpose, of course, is the best safeguard against boredom. But cultivating a variety of recreational interests and hobbies early in life can be an effective safeguard at all stages of living when boredom strikes.

Noise. In recent studies dealing with environmental pollution, much attention has been given to the role that noise plays in creating stress disorders. There is undeniable evidence that continuous exposure to loud noises greatly increases the production of adrenalin hormones with consequent harmful effects. In laboratory experiments, rats have been killed merely by subjecting them to the sustained stress of a screeching siren. The internal damage humans are subjected to from the noise of jets, trucks, cars, blaring and unattended TV sets and radios can only be surmised. Brain wave studies have shown that even mild nighttime noises can be stressful and disturbing. Even though not enough to waken a person, they can bring him to the threshhold of wakefulness and rob him of the benefits of deep sleep. The idea of the home as a place of "peace and quiet" has all but vanished in the midst of whirling kitchen appliances, fans, vacuum cleaners, and blaring stereos. Communication at the dinner table is hampered both by household noises and the general tenseness created by daily exposure to noise. In view of these damaging effects, much is to be said for the now almost forgotten practice of quiet pre-bedtime reading or, if possible, a quiet walk through a park or along a fairly deserted street.

Overcrowding. Most of us are also familiar with the emotional and physical stress of being caught in crowded buses, stores, restaurants, or the morning and evening highway rush hours. Overcrowding is another adrenalin producing situation that seems to bring out the worst in people: rudeness, hostility, and a variety of antisocial behaviors. Though admittedly difficult to escape completely, sometimes a change in schedule can be of some help. Leaving home a half hour earlier in the morning or leaving later from work to avoid peak rush hour loads is one way. Taking a longer, but less crowded route is another. Adjusting a lunch schedule to avoid crowded restaurants or cafeterias or bringing lunch from home is another minor adjustment that can bring major rewards.

Fatigue stress. True fatigue comes from the accumulation of metabolic waste products in overworked cells and results in lower efficiency, irritability, and other emotional stresses. For both physical and mental fatigue caused by overwork, adequate rest, of course, is needed to allow the body to rid itself of the accumulated waste products. But lack of sufficient activity and diversion, more so than overwork, is often a common cause of fatigue. The sluggish circulation and lack of glandular stimulation that results causes what is described as a "false fatigue"—tiredness, irritability, apathy—all the symptoms of true fatigue, but caused by "underactivity" rather than overwork.

Sedentary living patterns void of adequate exercise and stimulating recreational activities commonly result in a vicious cycle of false fatigue: the less one does, the more tired one gets; the less one feels like doing, and the less one does, etc. Breaking and reversing the cycle begins with instituting a regular program of exercise and recreational activity.

Even in well-conditioned individuals, however, a false fatigue can occur throughout the day from prolonged sitting, standing, or the monotony of performing repetitive tasks—no matter how light. Such times call for a short break and diversion to stimulate circulation and glandular activity: stretching, a short walk down the hall, a pause to look out the window—anything that stimulates circulatory activity or is different enough to break the stress of monotony.

**exercise, recreation,
and physical relaxation techniques**

Being aware of the damage caused by too much stress, asking in stress situations "if it's really worth it," examining causes and removing or reducing them where possible are some of the ways of lessening dangerous stress and tension. Other more specific methods include exercise, recreation, specific muscle relaxation techniques, and adequate rest.

exercise and tension

Vigorous exercise can lessen the detrimental effects of adrenalins produced by emotional stress in several ways. First, by distributing the stress over a wider area, the danger of disproportionate wear on one organ or system is reduced. Secondly, by strengthening the body and vital organs, they can better withstand the attacks of emotional stress which might otherwise bring about heart failure, other illness, or death. Finally, there are sufficient studies to show that the physiological effects of emotional stress can be "worked off" by vigorous exercise.

Scientific proof that exercise is effective in reducing tension has recently been made possible through a technique known as *electromyography,* the electronic measurement of activity in skeletal muscles. That a relationship exists between the degree of muscle tension and various psychologically induced states of anxiety and tension (nervousness, etc.) has long been recognized. EMG (electromyography) measurements have now enabled researchers to provide objective measurements of previously unmeasurable symptoms by relating the degree of electrical activity present in the muscles to various types and levels of stress.[2]

Various studies have shown that sampling one or two representative muscles in the resting state can provide good evidence on the state of the entire organism at any given moment. Further studies have been able

[2] Herbert A. de Vries, *Physiology of Exercise* (Dubuque, Iowa: William C. Brown Company, 1966), pp. 251–53.

to relate various EMG measurements to such conditions as headache, backache, mental activity, and various emotional states. Similar measurements have proven the effectiveness of exercise and other techniques in reducing residual neuromuscular tension.

The value of exercise in reducing tension is well recognized by doctors and psychiatrists. In a survey of over 400 California physicians conducted by Dr. O. E. Byrd of Stanford, it was found that exercise in the form of walking, golf, tennis, bowling, or calisthenics was recommended by most of these doctors as a means of combating nervous tension. A similar study among psychiatrists also showed that the great majority recommended physical activity for tension relief. Many others echo the sentiment that physical activity and sports, when an enjoyable part of everyday living patterns, serve—better than pills—as an excellent tranquilizer for the release of tensions.

recreation

Pleasurable, diverting recreational activities have also been recognized as an effective means of relieving harmful stress and tension. To better appreciate the significance of recreation, it should be thought of in terms of its true meaning: "to re-create," or to engage in activities that offer an opportunity for a person to re-create or renew himself. Unfortunately, many so-called recreational activities people engage in could be more aptly called "wreck-reational" activities.

Recreational activities can take many forms. They can be active or passive; creative (in a material sense) or noncreative. They can require varying degrees of physical, mental, or emotional participation. They include active participation in sports, games, or hobbies, as well as a variety of interests of a purely passive and vicarious nature, such as movies, books, or spectator sports. However, what is recreation for one person may be hard work and added stress and tension for another. In order to be successful in relieving stress and tension, the activity must be something that a person really enjoys and that he can get involved in to the point of forgetting troublesome problems or worries.

Generally speaking, recreation should involve some physical or mental activity that is separated from a person's normal work activities. This allows a different set of nerve structures to come into play, thus relaxing those that have already been stressed. A physically active person may find relaxation in reading a book or magazine article, but to an editor who reads all the time, this would be a continuation of the same stress he is subjected to all day. Similarly, to a person whose job involves a great deal of competitive stress, highly competitive sports activities would just add to his difficulties. Ideally, recreation should complement what a person normally does. The person whose job requires a great deal of interpersonal contact will usually find relief in individual-type activities while the one who is relatively isolated all day will find stimulation from group-type activities.

Experimenting with and cultivating skill and interest in a variety of recreational activities is also one of the most effective safeguards against mental and emotional illness. A survey conducted by Dr. W. C. Menninger of the Menninger Clinic which specializes in the treatment of mental and emotional illnesses showed that well-adjusted individuals have more hobbies and also participate more intensely in these activities than do the less well-adjusted. In explaining the value of recreational activities to the mentally healthy, Menninger said: "...their satisfaction from these activities meets deep-rooted psychological demands, quite beyond the superficial rationalization of enjoyment.... By comparison with two generations ago, there is today a greater need for recreative play. People now have little opportunity to express their aggressive needs, to pioneer, or to explore."[3] He cites as being of particular value the outlet that competitive games provide for the instinctive competitive drive and the psychological values that other forms of recreation provide: the opportunity to create, cater to passive desires, and gain vicarious fulfillment of creative and aggressive drives, such as from listening to music, studying art masterpieces, reading a mystery book, or seeing a ball game.

specific muscle relaxation techniques

Another way of relieving stress and tension is through certain techniques aimed specifically at reducing "muscular tension." In order to understand how these techniques work, a few basic facts about muscle tone (or tonus) should be reviewed.

Muscle tone is the term used to describe the slight state of muscular contraction that is normally present in the muscles of every healthy individual even when at rest. This tonus or tension increases when a muscle is consciously used or when a person is under psychological stress (a call to action). Other factors that play a part in the degree of muscle tone are poor health, pain, and temperature.

With regard to psychological stress, therefore, regardless of the source—worry, anxiety, fear, anger, frustration—one element is always present: "muscle tension." One cannot exist without the other. Get rid of the

3 From W. C. Menninger, "Recreation and Mental Health," *Recreation*, 42 (November 1948), pp. 340–46.

muscle tension, then, and the other elements will disappear. This relationship between the emotions and muscle tension was recognized many years ago by William James, the eminent psychiatrist from Harvard, who said, "If muscular contraction is removed from emotion, then no emotion is left." It is this concept that forms the basis for all the various techniques aimed at relieving emotional stress and tension by depressing muscular activity. These include massage, whirlpool, steam and sauna baths, and voluntary muscle relaxation.

Voluntary muscle relaxation. In 1929 Dr. Edmund Jacobson was one of the first to devise a specific training method for teaching a person to relax his muscles at will in moments of mental or emotional stress. The theory is that though we can't always eliminate the causes of tension, the muscles are always under our control. Therefore, if a person can learn to relax "controllable muscles," he can calm what appear to be "uncontrollable tensions." Thousands have gained a new serenity from developing this ability.

The method involves teaching a person to recognize and consciously release progressively decreasing levels of voluntary muscle contraction (tension) in various muscle groups until his perception is great enough to identify and remove the slightest degree of involuntary muscle tension from any source. The method was successfully used on Navy pilots during World War II and by students in various colleges and universities. One study on normal young subjects showed a reduction in neuromuscular tension of 25 to 40 percent (using EMG measurements) over an eight-week period, compared to no change in a control group which had not practiced the technique.[4]

Practicing the technique five or ten minutes a day may return you rich dividends. A good time to practice is before going to bed at night or in conjunction with a midday nap. The speed with which you learn to relax your muscles at will will depend on your degree of concentration and perception of muscle tension. Once the ability to recognize and remove muscle tightness is developed, it can be applied in any situation throughout the day when tenseness develops.

Jacobson's progressive relaxation training technique. The training technique involves the conscious contraction and relaxation of various muscle groups throughout the body at three levels of intensity executed in a progressive pattern. Though various patterns may be used, the normal order of progression is from the large muscles of the legs and then up through the gluteals, abdominals, back, chest, shoulders, arms, neck, and finally the smaller muscles of the face.

4 de Vries, *Physiology of Exercise,* p. 252.

The starting position for the exercises is on the back with the legs straight, arms at the sides, and eyes closed. Begin by slowly and deeply inhaling three or four times. Make your mind as nearly blank as possible and let yourself go limp. For each muscle group, the muscles are slowly contracted three times, either statically or isotonically, and then slowly relaxed until the muscle is in a limp, calm position, with no trace of tenseness. Each of the three contractions is made at a progressively lower level of intensity: first, maximum; second, approximately half strength; and last, very slight. The object is to concentrate on the "feeling" of muscle tenseness as it appears with each contraction and disappears when you consciously relax the muscle and allow it to go limp.

To get the initial feeling, sit or stand in a relaxed position with the arms hanging loosely from the sides. Let the shoulders "hang loose" on their bones. Notice that the fingers will be slightly curved. This is normal muscle tone. To straighten them would require a contraction of the forearm and create tenseness. Now tighten the fist and note the sensation—that's muscle tension! Now let the hand slowly relax until it is back into a fully limp position—that's relaxation; and the important thing is that it was controlled. If you apply the same principle in all of the following movements, you'll soon be able to recognize and remove tenseness in any muscle group at any time.

The detailed training procedure will be described only for the first training exercise. Only a description of the movements themselves will be given for the remaining exercises. The exercises should be performed in the order given and in the same general manner as the first one described.

1. Slowly *bend the right foot* toward the face as far as possible—feel the leg muscles tighten—and hold for a moment. Now slowly let the foot return to its normal position and go into a completely relaxed position for a few moments. Now bend the foot about halfway—feel the muscles tighten again. Hold and then slowly return the foot—feel the tenseness leaving—let the foot go limp. Feel the calmness in the leg and foot. Now just barely bend the foot and feel the very slight tenseness—remove it by letting the foot go again into a completely relaxed and limp position.

2. *Repeat with the left foot.*

3. Slowly *extend the right foot* away from the body as far as possible. Hold and return to the limp position. Repeat with half and then very slight movements.

4. *Repeat with the left foot.*

5. *Lift the right leg* up about 12 inches. Let it down and "let go," allowing the body to completely relax and the foot to just hang loosely to the sides. Repeat by lifting 6 inches and then with just the beginning of a

contraction that would start a lift but not move the heel from the ground.

6. *Repeat* with the *left leg.*

7. Statically *contract the gluteals* as hard as you can, and then let go. Let the rest of the body go at the same time. Repeat with half and then very slight contractions.

8. *Tighten the abdominal muscles* as hard as you can, and then let go, along with the rest of the body. Repeat with half and then very slight contractions.

9. Keeping the head, shoulders, and hips on the floor, *arch the back* as much as possible and then let go. Let the rest of the body go too. Repeat going halfway, then with just the beginning of a movement that would arch the back but not actually lift it.

10. Keeping the head and elbows in place, *tighten the chest and upper back muscles* by moving the shoulders inward as far as possible—return and let go. Repeat with half and slight movements, letting go completely each time.

11. Keeping the head and elbows in place, *contract the upper back muscles* by pressing the shoulders back against the floor—repeat with half and very slight movements.

12. Keeping the rest of the body relaxed, *clench the right fist* as hard as you can and then slowly let go. Repeat with half and very slight contractions, letting everything go with the arm.

13. *Repeat with the left fist.*

14. Now take a deeper breath than normal and *exhale slowly* until the chest sinks as far as it will go without forcing air out—let go completely. Take another deep breath and exhale—this time extending the exhalation phase even more. Let go completely: arms, chest, back, gluteals, abdominals, legs, and feet.

15. Keeping the rest of the body relaxed, *raise the head forward* and feel the tenseness in the front and back of the neck. Slowly return and let go. Then raise it just a few inches, and then merely begin to lift the head but keep it in place. Let go completely.

16. With the eyes closed and at three levels of intensity: full, half, and very slight:

   a. *Wrinkle the forehead*—hold it and slowly let go. Let the forehead and all the face muscles sag, but without forcing.
   b. *Close the eyes tightly* and let go.
   c. *Clench the teeth* and let go—let the jaws and other face muscles sag.
   d. *Open the jaws wide* and let go.
   e. *Move the cheeks back* and show the teeth. Let go.
   f. *Pucker the lips*—let the whole body go.

Applying the technique. By being on the lookout for muscle tenseness as it develops in various situations throughout the day, and practicing these techniques, you'll soon learn to recognize tightness in any muscle group and voluntarily relax these muscles. Be especially alert for telltale signs of tension, such as biting the nails, clenched teeth and fists, and tightness in the neck and shoulders and when you discover them—let go.

In an article entitled "How to Relax Without Pills," Leavitt Knight, Jr. offers these four further suggestions based on Jacobson's theory of muscle relaxation: "For one month, police yourself to habitually (a) keep your hands and arms limp when not in use; (b) keep your face— especially your lips and brows—placid when not talking and no more active than necessary when talking; (c) let your shoulders hang on their bones unless they have a load to bear; (d) let go any needless rigidity in your legs and feet when they aren't carrying you. If you do these four things, you will find a pleasant change toward serenity creeping up on you." In moments of pressure, excitement, hurry, and argument, he offers this further advice: "Notice the muscle tension that *always* goes with them. Now relax every one of those muscles that you can without falling down or looking ridiculous. Maintain all the muscular relaxation you can while still playing your role in the situation. Try it 40 times, not just once. Then you be the judge of what seems to happen to your part of the tenseness of the situation."[5]

stretching and pulling exercises

Some muscle tensions are the result of required set postures of work throughout the day, such as while writing, typing, driving, etc. A few simple pulling or stretching exercises are usually effective in relieving such muscle tension buildup.

1. *Shoulder and Head Pull:* Sitting or standing, place the arms behind the back with the left hand grasping the back of the right and pressed against the buttocks. Now slowly push the hands downward, pull the shoulders back, the shoulder blades down, and the head back as far as possible; then pull the shoulders forward, the shoulder blades up, and the head forward as far as possible. Repeat 2 or 3 times.

2. *Shoulder Shrug:* Sitting or standing with the arms at the sides, slowly pull the shoulders up and back as far as possible, and then let them drop into a completely relaxed and normal position.

5 From Leavitt J. Knight, Jr., "How to Relax Without Pills." Copyright *American Legion Magazine,* October, 1970. Reprinted by permission.

3. *Shoulder Roll:* Sitting or standing, with the arms at the sides, pull the shoulders forward and then back as far as possible, and then let them drop into a relaxed position. Repeat 2 or 3 times.

4. *Back Fold and Unfold:* Sit or stand, arms relaxed at the sides. Do two or three side bending and twisting movements first and then bend forward and gradually curl up into a ball with the head dropped, back rounded, and arms hanging loosely. Slowly return to the starting position by gradually straightening the spine—beginning at the lower back, and then the upper back, shoulders, chest, and finally the head until the return position is assumed. Repeat 2 or 3 times. *Note:* By doing this in a standing position, with the knees locked, tension in the back of the thighs can also be relieved.

5. *Foot Stretch and Pull:* Sit on the floor or edge of a chair, with the legs straight. Slowly curl the toes under and straighten the feet as far as they will go. Release and bring the feet back toward the face as far as possible, spreading the toes at the same time. Release and let the feet relax in their normal position. Repeat 2 or 3 times.

6. *Foot Roll:* In the same position, slowly rotate the feet, first in a clockwise direction, then counter clockwise, stretching them as far as possible throughout. Repeat 2 or 3 times.

yoga and breathing exercises

No discussion of relaxation would be complete without some mention of yoga—the ancient system of exercise developed in India which is gaining new popularity in America. Following is a brief description of yoga and some fundamental breathing exercises that can be easily learned, practiced, and added to your repertoire of relaxation techniques.

There are sixteen systems of yoga, but the one most concerned with improving health, revitalizing the body, and providing relaxation is *hatha yoga*. It is done by teaching various *asanas* (first syllable pronounced "ah"), or postures, together with breathing control. Though physical, hatha yoga is not confined to the body. Mental control, leading to relaxation, is practiced by focusing the mind on each of the various postures and meditating on their effects.

While doing asanas, there is a certain logic and sequence of the postures. The curving segments of the spine—at the neck, upper back, lower back, and base of the spine—are systematically flexed and strengthened. Every joint is mobilized and each ligament stretched. Every organ and gland is toned up through specifically applied pressure on certain body regions, promoting adequate circulation of blood throughout the body. Gradually, the system is conditioned to increasingly strenuous demands. Then comes a tapering off period with sitting postures which facilitate deep breathing. A final long relaxation period, while lying flat on the back, replenishes inner resources and allows time to assimilate the benefits of the asanas. For our purposes, we wish to cover only some of the basic breathing exercises which can aid relaxation.

Breathing exercises and relaxation. Deep breathing has a generally calming effect on the body and whole personality, reducing jumpiness and nervous tension. The creatures that breathe more rapidly than man tend to be more restless and short-lived. Monkeys have almost twice the breathing rate of man, whereas the long-lived elephant and tortoise are slow breathers. One thing the yoga student learns to do in moments of stress—or in rallying energy—is to deliberately slow down his rate of breathing. Nervousness decreases and composure is induced.

Correct breathing technique. Correct breathing is performed from the diaphragm, with both the stomach and chest expanding during inhalation and receding during expiration. Animals, as well as children and primitive people, breathe naturally from the diaphragm, relaxing the stomach—as do all persons during a relaxed sleep when the system must refresh itself. Many people, however, especially when nervous or tense, will restrict or tighten the stomach while breathing, rather than relax it. This contributes to a faster rate and shallow breathing. Though it is not necessary or desirable to be constantly thinking about how you breathe, correct breathing technique can become automatic by practicing a few simple breathing exercises. There is nothing unnatural and everything natural about them. They teach a person to breathe as nature intended, enlarging lung capacity and expelling old stale air in addition to aiding relaxation. They have helped some persons to quit smoking, merely by substituting deep diaphragmic breathing at the times they would normally light up.

Simple diaphragmic breathing exercise. To develop the feel of correct diaphragmic breathing and its relaxing effect, start with the following exercise:

1. Lie flat on the back with the knees pulled up and feet slightly apart. Inhale deeply and allow the stomach to relax and expand like an inflated balloon. Exhale and pull the stomach in so that the diaphragm rises and presses against the rib cage. Continue until the rhythm seems easy and natural.
2. Continue to breathe diaphragmically, but now after filling the lower portion of the lungs, concentrate on filling the middle and upper portions as well. Note the difference in capacity. Observe also how the air seems to fill up in a wave-like motion.
3. Exhale, visualizing the air pressing out first from the lower then the upper chest cavity. Observe the same wave-like motion of the air as it passes out of the body.

4. Continue to breathe slowly and deeply, concentrating on the wave-like motion of the air as it enters and leaves the lungs, until the breathing is very slow and regular and the body completely relaxed.

Breathing exercise—sitting. This simple breathing exercise was developed by William P. Knowles, Head of the Institute of Breathing in England and has helped many in their efforts to give up smoking.

1. Sit with your hands on your thighs, elbows at the sides, and spine straight, but without touching the back of the chair.
2. Breathe in and out rapidly through the nose, similar to the rhythm of sawing wood, about 12 times.
3. Now slowly exhale through the nose until there is no air in the lungs.
4. Inhale to the count of seven; pause one second, and exhale again. Pause one second and repeat for a total of seven complete breathing cycles.

rest

A final weapon in the arsenal against stress and tension, as well as against fatigue, which can conserve and restore energy and prevent tissue breakdown is adequate rest. Both quiet sitting or lying, with or without casual conversation, are valuable forms of rest throughout the day. But of special importance is that special form of rest we call sleep.

Sleep is one of the most neglected and abused areas of living for many people. Though a person would not think of going without adequate food for a long period of time or without proper clothing in cold weather, going without adequate sleep often doesn't seem as important. Yet it remains one of the basic essentials of life.

All of nature seems to have a built-in "biological clock" which calls for such a form of rest for a long interval during each twenty-four hour solar day. In man sleep was once looked upon as a completely unconscious state, associated with a form of "temporary death." Perhaps this explains why some people try to get by with as little as possible. But recent studies on the nature and importance of sleep show clearly its unique role for the health of both the mind and body.

nature and importance of sleep

Brain wave studies have shown that rather than being a total state of unconsciousness, sleep involves a recurring cycle of deep slumber with minimal brain and bodily activity and periods of an intense inward experience characterized by dreams, many kinds of thought, and physiologic activity closely resembling the waking state. These alternating phases of "deep sleep" and "light sleep" repeat themselves several times during the sleep period, changing only the relative amount of time spent in each phase. The deep sleep phase, called Delta sleep, predominates in the early half. In the latter half, and especially as the end of the sleep period nears, the lighter phase called Alpha sleep predominates. Since it is during the lighter stages of the Alpha phase that dreams occur, detected by rapid eye movements (REM), this phase is also referred to as "dream sleep" or "REM sleep." Delta sleep, on the other hand, is dreamless.

Though it is difficult to waken a person from deep Delta sleep, during the light Alpha phase he approaches the threshold of wakefulness and can be awakened at the slightest sound or movement. Whether a person awakens feeling fresh or tired quite often depends on what portion of the cycle he was caught in when awakened. If just prior to the short Delta period, he will usually feel tired. This is when that little extra "cat-nap" seems most inviting and—very likely—is necessary and advisable.

Importance of dream and deep sleep. Normally, it appears that every-one dreams about 20 percent of the time, with the dreams progressively increasing in length and vividness as the Alpha sleep period becomes longer. If a dream is recalled, it will most likely be the most recent REM dream and one that lasted about thirty minutes.

Psychiatrists believe that dream sleep serves as a valuable safety valve for the mind. It is here that the day's frustrations are thrashed out and wishful thinking takes place. In studies of the mentally ill, little REM dream time has been observed. In other experimental studies, persons deprived of dream sleep showed obvious deleterious changes in behavior, including anxiousness, irritability, and inability to concentrate.

Several years ago, a disc jockey with some record of mental instability went seven days without sleep and ended up in a mental hospital. In every instance where persons have gone for long periods without sleep, either forced as in military combat or for publicity purposes, a similar pattern of psychotic-like symptoms has occurred, including a lost sense of time, confused thought, delusions, lapses into fantasy, and vivid auditory and visual hallucinations.

After adequate rest, most have recovered without any noticeable traces, but laboratory tests have shown that prolonged sleep loss causes sizable changes in body chemistry and brain function. Whether the changes can lead to irreversible damage is not yet known. Animal studies have shown noticeable changes in brain tissue. For humans, it would be safer to err on the side of caution and guess that prolonged sleep loss would leave some permanent damage.

It should be noted also that continuous irregular or interrupted sleep from any cause can bring on the same neurotic and psychotic symptoms as prolonged sleep loss, since when sleep is abbreviated, the primary effect is loss of dream time. Hence, not only does abbreviated sleep change the quantity but the quality of sleep as well.

Though the brain still remains open to certain kinds of stimuli, deep sleep is dreamless, but its importance should be obvious. It is here that the body and brain activity reach the lowest level of the day. Accumulated waste products in both muscle and nerve cells are able to be removed at a faster rate than they accumulate and the cells are able to repair and rebuild themselves. Worn processes are rebuilt and energy restored for the next period of activity. Fatigue and inefficiency are the obvious results of inadequate deep sleep. Studies again have shown that when deprived of this stage, the body will automatically compensate for it by longer periods on succeeding nights with a proportionate decrease in dream time.

effect of drugs on sleep

Though over-the-counter sleep inducing drugs seem to have no effect on sleep patterns, large doses of barbituates and alcohol are known to interfere drastically with dream sleep, making it difficult if not impossible to induce. The primary effect of over-the-counter aids appears to be merely pyschological. There is danger, however, of becoming "psychologically hooked" on these otherwise harmless aids to the point that nightmares occur when stopped. Experience has shown that, better than any artificial aid, the relaxing effect of exercise and the other relaxation techniques presented in this chapter have helped many to gain the ability to sleep well.

how much sleep is needed?

There is no blanket formula for sleep needs that can be prescribed for all, either in terms of total amount or specific time periods for sleeping. The only certainty is that all require a certain amount of dream and deep sleep during each twenty-four hour period. How much and when it should be taken varies with each individual depending on such factors as type of life, physical condition, amount of physical or mental activity, tension, ability to relax, and many others including temperature cycles, levels of metabolic activity, and personality characteristics. For most persons, the normal range seems to be six to eight hours a day, but studies have shown that some persons felt more refreshed with two separate three hour rest periods than after an eight hour sleep. The best way to decide if you have slept enough is to decide if you feel rested when you awake.

Personality characteristics and metabolic activity, however, apparently are closely related to both the amount of sleep required and the time it is most needed.

Personality characteristics. According to Dr. Earnest L. Hartman, Director of the Sleep and Dream Laboratory at Tufts University, short and long sleepers show strikingly different personality characteristics. Those who sleep less than six hours a day are usually energetic, outgoing, efficient, and achievement-oriented. They are frequently "turned on" in speech, thought, and motion. Those who sleep nine hours or longer, on the other hand, are usually introverted, passive, slow-moving, and subject to mild depressions. Studies showed that both types spent about the same amount of time in deep sleep, but that the long sleepers had two times more dream sleep. Dr. Hartman's conclusion was that, regardless of activity level, all seem to require the same amount of deep sleep, but the amount of dream sleep needed depends on personality characteristics. In addition to presenting a general guide for sleep needs, a possible implication for long sleepers is that if they have never tried getting along on less sleep, they might try it for a week and see what happens. It may open up a new life and new personality.

Metabolic activity patterns. In terms of metabolic activity patterns, Dr. S. B. Whitehead in *Facts on Sleep* classifies people as having one of three predominant rhythms. The "morning type" awakens raring to go with a metabolic activity in high gear. By mid-afternoon, it begins to slow down and by the evening, it comes to a dead stop. While he may be ready to call it a day at 9:00 P.M., the "evening type" is just hitting full stride. He's the one who drags out of bed in the morning with a metabolism in low gear. He eases through the morning, but by afternoon when others are starting to fade, he suddenly becomes a "spark plug." By evening, he's in full stride. The third type is considered the most fortunate. This is the person whose metabolism is such that he hits full stride in both morning and evening, suffering only a moderate mid-day letdown.

Along with determining the optimum amount of sleep you need, determining your particular metabolic rhythm can help you adjust your sleeping habits to a pattern best suited for you. "Early to bed, early to rise" may be the best pattern if you are the A.M. type, but if you are the P.M. type, a "late bed, late rise" pattern (if possible) may help you function more effectively. Knowing your particular rhythm can also help you make effective use of "naps" which can serve as valuable energy rechargers at anytime—especially at low points. For example, if you are the A.M.–P.M. type, a half hour nap after lunch can be as effective as the last three hours of sleep just before waking. For the morning type, a thirty minute nap in mid-afternoon or before dinner can work a miracle in preparing for a late evening special task or social engagement.

Whether for fifteen minutes, thirty minutes, or eight hours, however, how well you rest is more important than how long. Regular exercise, mastery of the muscle relaxation techniques presented, and mind diverting recreational activities can improve the quality of your rest and with it your ability to handle stress and tension.

9

sport and recreational activities for fitness

A final component of fitness is knowledge and skill in a wide variety of sport and recreational activities that can be used as wholesome leisure-time activities to provide fun and relaxation, and as valuable supplements in the overall fitness program. There are many opportunities available to learn these activities—schools and colleges, city recreation departments, and other public and private organizations and clubs. To make the most of them, however, selection of these activities should be carefully made and you should have some idea of what is involved in learning sports skills.

Many times individuals choose an activity on the basis of little more than a vague interest, influenced perhaps by the urging of friends. Little attention is given to the values of the activity in terms of meeting present and future needs. Some are afraid to try an unfamiliar activity out of fear they will not succeed.

In order to help you make a more intelligent selection of activities, we will present in this chapter certain factors that should be considered. We will conclude with a discussion of certain basic principles involved in learning sports activities which can help make learning easier and more effective.

considerations in selecting activities

immediate physical needs or limitations

Of first consideration is your present level of fitness. If you are seriously overweight, physically underdeveloped, or lacking in fundamental motor skills (running, jumping, throwing, basic body coordination, etc.), the first step is an intense program specifically aimed at improving these areas. This would include any program in the strengthening and conditioning area shown in Table 10 (jogging, calisthenics, weight training, etc.), or perhaps an individual adaptation program where your specific needs can be determined and an individual program worked out under supervision. Most colleges and universities offer such a program.

Should you have some temporary or permanent disability, professional advice or enrollment in an individual adaptation program is definitely needed. You can then be guided into those activities appropriate to your ability or into a planned and supervised program for correction if possible.

ability to swim

After considering your immediate physical needs or limitations, the next consideration is your swimming competency. The ability to swim and familiarity with the basic skills of water survival is a must for anyone. It has been estimated that drowning accounts for 6 percent of all accidental deaths. Many of these drownings could have been prevented had the victim or those with him been familiar with basic water survival skills. Often the simple ability to stay afloat or the calm extension of a supporting object is all that would have been needed to save a life. But because of ignorance or panic stemming from inexperience, the opportunity was lost. In addition to being a potential lifesaving activity, the ability to swim can open up a host of pleasurable recreational activities that otherwise would go untried: sailing, boating, fishing, skin and scuba diving, water skiing, and others.

If your present level of condition is satisfactory and you are basically secure around water, you can then turn your attention to other activities, giving consideration to the following factors.

carry-over value

For college students, the carry-over value of an activity—or the feasibility of continuing it throughout life—is an important consideration. Fitness and recreation are lifetime concerns; they don't end with graduation. Unless an activity can be performed well into adult life, its benefits will be short-lived. Many of the highly popular team sports such as football, baseball, and basketball fall into this category. To concentrate on these types of activities, as popular and appealing as they may be at the moment, would be a mistake.

Certainly the overwhelming majority of college students have had ample experience in team sports throughout their elementary and high school years. Unless playing on an intercollegiate team, there is little more than can be learned about a major team sport—and even less need for it. Furthermore, the desire to participate in team sports can be satisfied in either the intramural, extramural, or intercollegiate programs without sacrificing the opportunity to learn new activities in the basic instructional program which can have more lasting physical, emotional, or social value. Many highly skilled athletes, often excused from physical education during their college playing careers, find themselves handicapped socially when they graduate. They find most of their new associates taking part in activities which they have ignored—perhaps feeling at the time that they were not demanding enough. The result is that they must then seek private lessons to learn activities they have missed out on. Since many of these activities (golf, tennis, archery, hunting, fishing) demand fine motor skills and coordination, they are usually more difficult to learn at a later age and discouragement is more common.

variety and balance

For all persons, consideration should be given to selecting activities that will provide enough variety to insure year-round participation and a balance between meeting various physical, social, and emotional needs.

Insuring year-round participation. Because of variations in climate and the seasonal nature of many sports activities, few can be engaged in on a regular, year-round basis. It is advisable, therefore, to have in your repertoire some activities that can be played outdoors and others that can be conducted indoors, so that climate cannot interfere with regular participation. Choosing activities that meet changes in seasonal popularity will also insure steady, year-round recreational activity. In addition, consideration should be given to the number of persons required for a particular activity. Depending solely on activities that require a large number of persons may restrict your opportunities. Include, therefore, some activities in which only one or two persons are needed for participation.

Meeting various physical needs. Each activity tends to contribute in varying degrees to the many components of physical fitness. Some place more stress on circulorespiratory endurance. Others may be more beneficial for muscular tone, strength, and endurance of various body parts or for agility, coordination, or flexibility. Some activities are excellent for those desiring to lose or control weight, while others are good posture aids. Though specific training in one or more of the exercise programs presented is best for development of these various components, being able to substitute more pleasurable sports activities during the maintenance phase will help to avoid boredom and make you more likely to continue a regular program of physical activity. Of particular importance in this respect are the aerobic activities which emphasize cardiovascular fitness.

Social needs. The social aspects of an activity are the opportunities it provides for making new acquaintances, deepening friendships through sharing mutually enjoyable experiences, meeting the basic need for recognition, and developing a feeling of belonging and security in group relations. Though most activities by their nature make a social contribution, some are more valuable than others in this respect. These are the activities that provide an opportunity to participate on an equal basis with friends, family, or others, regardless of age, sex, or skill level, and which allow warm, interpersonal communication during the activity. They include certain competitive activities in which competition is equalized by a handicap system (golf, bowling) or those that can be paced without losing their enjoyment (badminton, tennis). Noncompetitive, purely recreational activities such as fishing, skating, and social dancing are also high in social values.

Emotional aspects. The emotional values of an activity are largely a matter of individual interest and temperament. What may be emotionally satisfying for one person may have no meaning at all for another. The characteristics of various activities, then, should be considered in terms of your individual temperament.

Some persons prefer activities which are competitive in a personal way rather than against another person; that is, they have a built-in challenge, such as archery, golf, or bowling. Some enjoy activities that are mentally as well as physically stimulating—activities that are quick moving and call for mental acuity. By *mental acuity* we mean quick decisionmaking and the application of strategy under physical and emotional stress, such as in handball or tennis. Others enjoy activities which offer a mental challenge, but which allow for slower, more deliberate thinking and analysis, such as in golf or bowling. Still others find greater emotional satisfaction from activities that allow for mental relaxation and little conscious attention to the performance itself (skin diving, fishing).

Occupation and residence. Two long-range factors to consider in selecting activities are your present or expected occupation and place of residence. If your occupational choice is a rather sedentary one, you should select more vigorous activities. If social contacts are important in the job, the more socially oriented activities could prove beneficial. These can also be valuable as a balance if the job is one in which social contacts are minimal, such as in research or as a technician. Balancing job activities that may be either mentally dulling or stimulating with recreational activities offering the opposite is another consideration.

With the nature of society as it is today, chances are that you will live in several geographic locations throughout your lifetime. Adapting your activity selections to the community recreational resources and facilities in the area and to the activities that are most popular there can help to make any transition easier as well as assure continuance of your fitness pursuits.

potential for success

The potential values of an activity should be weighed against potential for success in the activity before making a final selection. Unless you will be able to perform the activity with reasonable success, the values will not be gained. Specific factors that will determine your potential for success are your body type, your general level of condition, specific motor fitness attributes in comparison with the requirements of the activity, and your experience in related activities. Related activities are those which share common skills. A skill learned in one related activity can be transferred to another and make learning it quicker and easier. All of these factors will have a bearing on the rate of progress and level of success you are likely to achieve. This in turn will influence your enjoyment of the activity and the likelihood of continuing it.

It should be pointed out, however, that "success" is more or less a relative term and depends on your particular goals or aspirations. Though your body type, general motor fitness abilities, and previous experience may make it unlikely that you will ever reach an advanced level of skill in an activity, you may not particularly care about becoming an "expert" or "good" performer. You may be perfectly satisfied with the values you gain at the beginning level, even if it takes you longer to arrive at them than someone else who may have more "natural" ability. In addition, the extent of your interest and motivation will have a large bearing on your ability to overcome certain shortcomings. Recognizing what these shortcomings may be, however, will prepare you psychologically for the difficulties you may face in learning and developing skill in the activity.

Body type. As pointed out previously, your basic body type will determine in part those activities you are more likely to have success in and
those that might be more difficult. Persons with at least an average
degree of mesomorphic components and medials will usually be successful in most activities at the beginning or intermediate levels. Those high
in endomorphy will find difficulty in activities which require rapid
movement, agility, or support of the body weight. They can be quite
successful, however, in such activities as swimming, golf, archery, bowling, skating, dancing, hunting, and fishing—where the physical requirements are less demanding. Ectomorphs should basically avoid sports
involving body contact. In terms of carry-over activities, they can probably be successful in almost any activity provided they have developed
an average degree of motor fitness. In its absence they may find some
difficulty in certain activities requiring strength in the arm and shoulder
girdle, speed, and agility. Generally, however, these physical attributes
can be developed through proper conditioning.

Present level of condition and specific motor fitness attributes. Your
rate of progress and potential for success will also be affected by your
present level of condition and specific level of ability in certain elements
of motor fitness in comparison with the requirements for a particular
activity. The requirements for success in an activity are essentially at
least to an average degree the same elements that the activity helps to
develop or maintain. For example, handball or racket sports can contribute to endurance, agility, coordination, flexibility, timing, judgment, and
a certain amount of strength and endurance in the arms, shoulders, and
legs. But before they can be played with a reasonable amount of success
to bring about benefits, at least an average level of ability in these areas
is required. If you are low in some of the key requirements of an activity
and are relying on the activity itself to develop them, you can expect slow
progress towards your ultimate level of achievement.

Before entering an activity for which you may be poorly equipped
(overweight, physically underdeveloped, or low in a key area), it would
be better to prepare yourself through an intensive training program in
the specific elements involved. At the least supplementing the activity
with a specific training program would be advisable. Once you have
reached an adequate level for performance, the activity can be used
for further improvement or maintenance.

For the development of some elements, however, such as timing,
judgment, agility, and coordination (including eye-hand coordination),
an activity may be the most effective training method. When used for
this purpose, consideration should be given to the difficulty or complexity of the skills involved. In some cases, a less demanding, easier

to learn activity can be used as a stepping stone to a more complex and difficult one that you may be more interested in, but not ready for. For example, one-wall handball played with a standard rubber ball can lead naturally to paddleball, and then to tennis. Each of these activities requires agility, similar positioning of the body to strike a moving object, and eye-hand coordination. But the progression from the relatively simple act of striking a standard-size ball with the open hand (a natural part of the body), to striking with a short paddle (an unnatural object) would allow for the gradual development of eye-hand coordination and related skills and make the introduction to tennis easier. Badminton is also easier to learn than tennis because of easier racket control, the smaller area of coverage, and the more easily adjustable pace which allows speedier progress to rallying and play. Because of this, it can also be used as a good beginning activity for the development of eye-hand coordination, agility, etc., and progression to tennis or squash.

Related activity experiences. As indicated, many skills developed through participation in one activity can be effectively carried over to another activity and make the learning easier and quicker. The relationship between handball, paddleball, tennis, and squash has already been noted. Previous experience and success in one of these related activities will increase your likelihood of success in another. Conversely, lack of success in a related activity area calling for similar skills would lessen your chances of success.

Other examples of related activities having basic skills in common are roller skating, ice skating, and skiing. The major common element in each of these activities is balance. Tumbling, trampoline, and springboard diving are others. Similarly, the same feigning, agility, and ball catching skills that a good basketball player has can be carried over as a wide receiver or defensive back in football. In selecting activities in terms of potential for success it is wise to look for activities related to those in which you have had some previous experience and success.

the activity selection guide

The Activity Selection Guide (Table 10) is intended to summarize some of the main points presented in this section and assist you in choosing those activities which best meet the criteria given for selection. By comparing the major contributions of the various activities with an evaluation of your personal needs, potential for success, goals, and level of interest and motivation, you should be able to narrow your choices to those which will be most valuable to you.

table 10

activity selection guide*

Activities are grouped according to recommended variety and balance. Beginning competency should be sought in each major area and subarea, with swimming a must.

Code
+ Good
− Fair
○ Low

Activities	Physical Values — CV Endurance	Agility & Balance	Coordination	Muscle tone, strength, & endurance — Arms–Sh'ld	Legs	Over-all	Emotional Values (Fun, Pleasure, Mood Adjust., Mental Relax., Achievement Recog., Indiv. & Group)						Soc. Val.	Carry Over Value	Present Competency (See Direction #1) / Beg. Adv.	Suitability for Pers. Needs (Physical, Emotional, Social, Fun, Fellowship, Indiv. & Group)	Suitability for Body Type — Endomorph	Mesomorph	Ectomorph	Medial	Choices (Follow Directions 2–7) / Ck. Rate
Indiv. Adaptation/Correct.	+	+	+	+	+	+	+	+	+	+	+	+					+	+	+	+	
I. Team Sports																					
A. Indoor																					
Basketball	+	+	+	−	+	+	○	○	−	−	−	−	−	−			○	+	+	+	
Ice Hockey	+	+	+	−	+	+	○	○	−	−	−	−	−	○			○	+	−	−	
Volleyball (Reg.)	−	−	+	−	−	+	○	−	−	−	−	−	+	+			−	+	−	+	
B. Outdoor																					
La Crosse	+	+	+	+	+	+	○	○	+	−	−	−	−	○			○	+	−	−	
Soccer	+	+	+	○	+	+	+	○	−	−	−	−	−	○			○	−	+	+	
Tag Football	−	−	+	−	+	+	−	−	−	−	−	−	−	○			−	+	+	+	
Softball	○	−	+	−	−	+	−	○	○	−	−	−	−	−			−	+	+	+	
II. Individual (Competitive)																					
A. Indoor																					
Badminton	+	+	+	−	+	+	+	○	+	−	+	+	+	+			○	+	+	+	
Bowling	○	−	−	○	+	+	○	−	○	+	−	−	+	+			+	+	+	+	

230

Activity															
Handball, Paddleball, Squash	○	+	+	+	+	+			+	+	+		+	+	○	+
Volleyball (2–3 man)	○	○	+	+	+	+			+	+	+		+	−	+	+
B. Outdoor																
Archery	+	+	+	+	+	+			+	+	+	−	+	+	+	○
Golf	+	+	+	+	+	+			+	+	+	+	−	+	+	−
Tennis	+	○	+	+	○	+			+	+	+	+	+	+	−	+
III. Individual-Group (Rec)																
A. Indoor																
Bridge, cards	○	○	○	+	+	+			○	○	○	−	○	○	○	○
Billiards/Pool	○	○	○	+	+	+			○	−	○	○	−	○	○	○
Chess	○	○	○	+	+	+			○	○	○	+	○	○	○	+
Ice Skating, Roller Skating	−	+	+	+	+	+			+	+	○	+	+	+	−	+
B. Outdoor																
Camping	+	+	+	+	+	+			○	○	○	○	○	+	○	○
Cycling	+	+	+	+	+	+			+	−	+	+	+	−	+	+
Fishing	+	+	+	+	+	+			○	○	○	○	−	○	○	○
Hiking	+	+	+	+	+	+			+	○	+	○	○	+	−	+
Hunting	+	+	+	+	+	+			−	○	○	−	−	○	−	−
Riding (Horseback)	+	+	+	+	+	+			○	+	○	−	○	+	−	○
IV. Combatives																
Fencing	○	+	+	+	+	+			+	+	+	+	−	+	−	+
Judo, Karate	−	+	○	+	+	+			+	+	+	+	+	+	+	+
Wrestling	−	+	○	+	+	+			+	+	+	+	+	+	+	+
V. Conditioning/Gymnast.																
Calisthenics	+	+	+	+	+	+			+	+	+	+	−	−	−	−
Gymnastics (Apparatus)	+	○	○	+	+	+			○	−	○	+	−	+	−	−
Gymnastics (Trampoline)	+	+	○	−	+	+			−	+	+	+	−	−	−	−
Gymnastics (Tumbling)	+	○	○	+	+	+			+	+	+	+	+	○	+	−
Jogging, Running	+	+	+	+	+	+			+	○	○	○	○	○	+	+
Weight Training	+	+	+	+	+	+			+	+	+	+	+	+	−	−

* Wayne D. Van Huss, Roy K. Niemeyer, Herbert W. Olson, and John A. Friedrich, *Physical Activity in Modern Living*, 2nd ed., © 1969. Adapted by permission of Prentice-Hall, Inc., Englewood Cliffs, N.J.

table 10 (cont.)

Activities are grouped according to recommended variety and balance. Beginning competency should be sought in each major area and subarea, with swimming a must.

Code
+ Good
− Fair
O Low

Activities	Physical Values — Muscle tone, strength, & endurance					Emotional Values			Soc. Val.	Carry Over Value	Present Competency (See Direction #1)	Suitability for Pers. Needs	Suitability for Body Type				Choices (Follow Directions 2–7)
	CV-End	ACF (agility, coordination, flexibility & Balance)	Arms–Sh'ld	Legs	Over-all	WP	MA	MI			Beg. Adv.	PE SF FI	EM	EM	EM	EM	Ck. Rate
VI. Dance																	
Contemporary	−	−	O	−	−	O	O	O	+	+			+	+	+	+	
Folk-Square	+	+	O	+	+	+	+	+	+	+			−	+	+	+	
Modern	−	−	O	+	+	+	+	+	+	+			−	+	+	+	
Social (Ballroom)	O	O	O	−	−	O	O	O	+	+			+	+	+	+	
VII. Aquatics																	
Beginning Swimming	−	O	−	−	−	O	−	+	−	+			+	+	+	+	
Intmd. Swimming	+	O	+	+	+	+	+	+	−	+			+	+	+	+	
Life Saving, W.S.I.	+	+	+	+	+	+	O	+	−	+			+	+	+	+	
Skin/Scuba Diving	+	O	+	+	+	+	+	+	−	+			−	+	+	+	
Surfing	+	+	+	+	+	−	−	−	−	+			−	+	−	+	
Water Skiing	−	O	+	−	−	+	+	+	−	+			−	+	+	+	
Boating	O	O	O	O	O	O	O	O	−	+			+	+	+	+	
Canoeing, Rowing	−	O	+	+	O	−	−	−	+	+			+	+	+	+	
Rowing	−	O	+	+	−	−	−	−	−	+			+	−	+	+	
Sailing	O	O	−	O	O	+	+	+	−	+			+	+	−	+	

232

The activities are grouped for variety and balance according to the categories shown on the Activity Competency Questionnaire which you completed during the fitness evaluation. Based on your present level of competence, you should attempt to balance out your choices so that you will have at least beginning level competence in each major area and subarea listed, *excluding* team sports and combatives. Once this is achieved, you can then either broaden your base by learning additional activities from any area or deepen your knowledge and skill in a particular activity through advanced programs, study, or practice. In this regard, it is advisable to try and develop a high level of skill in at least one activity.

explanation of chart ratings

Activities are rated Good ($+$), Fair ($-$), or low (0) in terms of their various physical, emotional, and social values, carry-over value, and suitability for predominant body types. Under muscle tone, strength, and endurance, the specific component(s) can be judged from the nature of the activity. Furthermore, it should be remembered that in terms of development, though two activities may have an equally high rating, activities in which resistance or speed overload can be easily controlled would be better. Emotional and social values, of course, are relative in many cases. The ratings given are approximations based on the average person.

directions

1. Start in the "Present Competency" column by placing a check mark for those activities· in which you have beginning or advanced competency. Where multiple activities are listed on the same line, circle the appropriate activity.
2. If warranted by your condition, check individual adaptation on the line in the "Choice" column and give it a rating of 1. Future choices will have to be made in consultation with your program supervisor.
3. If you are not in need of an individual adaptation program, but do need more intensive work in the strengthening and conditioning area, check an appropriate activity in area V and rate it 1.
4. The next consideration is your ability to swim. If you do not know how to swim, check "Beginning Swimming" in the "Choice" column and give it a rating of 1 or 2, depending on whether or not you have checked the strengthening and conditioning area.
5. Now for each subgrouping of activities in areas II, III, and VI in which you *do not* have at least beginning level competency, examine

the values listed for each activity and in the "Suitability for Personal Needs" column make an appropriate rating for each.

6. After making this evaluation, weigh each against your potential for success in the activity (suitability for body type, related experiences, motivation) and then, in the "Choice" column check one activity from each area. If your interest is high enough, do not be discouraged from choosing an activity even if you feel your physical potential for success is low. Potential for success is only in terms of the ease of learning, rate of progress, and chances of developing a high level of skill. With enough motivation, you'll still be able to develop enough skill to take part in the activity, even if at a relatively low level. If you can be satisfied with this possible limitation and still see other high values coming from it, then go ahead.

7. Now, according to need or preference, rate the activities checked in the order you wish to learn them, starting with either 1, 2, or 3, depending on previous activities rated.

When you have completed this activity learning program, you will have a minimum but varied and well-balanced arsenal of activity skills to meet your future needs. Though not as critical in making choices above the minimum level, you should again try to weigh the potential values against your needs and potential for success.

guidelines for learning sports activities

There is really no such thing as "general motor ability" that is transferable to all physical activities. Each activity calls for specific physical skills which must be learned. Certainly inherited characteristics, a high level of motor fitness, and past experience in related activities can make the learning process easier, but the specific skills must still be learned. An awareness of certain principles and characteristics of motor skill learning can be of great help to you in learning various sports activities.

motivation

Motivation is the heart of any learning process. Without the desire to learn, learning is almost impossible. The degree of your motivation will determine not only the extent of your needed personal involvement in the learning process, but how persistent you will be in overcoming the difficulties and frustrations that will inevitably appear. Within your inherited potential, motivation will be the major factor determining your rate of learning and improvement.

Essentially, your motivation will depend on your purpose or motive for learning the activity. Unless you see learning the activity as a means of fulfilling some need or desire, motivation will be absent. The nature and extent of the need or desire will also determine the *degree* of your motivation. It will be tempered by the confidence you have in your ability to learn, your previous experience of satisfaction or disappointment in motor activities, and your general attitude towards learning. Your instructor's attitude, encouragement, and teaching methods can help to identify values, develop confidence, and inspire the desire to learn, but the basic drive must come from you.

Intrinsic and extrinsic motivation. Motivations for learning are generally classified as intrinsic or extrinsic. *Intrinsic motivation* refers to values seen as being derived from performance of the activity itself. In other words, your desire to learn and participate comes from enjoyment of activity for its own sake. Fun and relaxation, physical benefits, a means of expression, a stimulating challenge, self-satisfaction from achievement, or the fulfillment of a personal drive for excellence are examples of intrinsic motivation. It is the best form of motivation.

Extrinsic motivation, on the other hand, is external to the activity itself, though still intrinsic to the performer. It may be negative or positive. An example of a negative extrinsic motive or incentive for learning would be to avoid being ridiculed for not knowing the activity. Positive extrinsic motivations come from seeing the activity as a means of gaining something external to that coming from the participation itself. It may be recognition, prestige, or the approval of parents, friends, or peers. For some, the motivation for learning and improvement may be to gain an award in the form of a certificate, letter, or trophy. Others may seek money or fame from the development of a high degree of excellence in an activity. Since these types of motivations are manifestations of basic drives, they can be considered partly intrinsic—provided they are not dominant to the point of unconcern for the inherent values of performance.

Intrinsic motivation is superior to extrinsic motivation and should be sought and worked for. In most cases, though, it is something that must grow with experience and a certain measure of success. Once achieved, an intense liking for the activity is developed and this provides the highest form of motivation for continued improvement and participation. Until such time, some form of extrinsic motivation is often a necessary starting point.

patience

One thing you will need in learning an activity is patience. Though many activities (tennis, golf) can be played after a brief introduction to the equipment, rules, and basic procedure for play, enjoyment of the activity is short-lived unless the essential skills of the game are adequately developed. This often requires breaking down the activity skills into component parts, working on each skill separately for a time, and then putting them all together within the total context of the activity. Other activities, such as gymnastics and swimming, by their very nature, require a progressive buildup of separate skills before the entire activity can be performed.

Some persons become impatient with this "part" method of learning and are overly anxious to "play the game" or perform the activity in its entirety. When learning an activity, however, concentration on just playing the game does not offer enough opportunity for the repetition essential for the development of skilled movements. Furthermore, emotional involvement in the activity, especially if it is competitive, will often lead to the sacrifice of sound and correct movement patterns to those which are wrong, though more effective for the moment. This practice sets a definite limit on the level of success and range of play opportunities than can be reached.

It is only through the repetition of correct movement patterns that the motor concept and "feel" of a correct movement *is* developed. For some with keen powers of observation, analysis, and "kinesthetic sense" (muscle sense), a skilled movement pattern is quickly developed. For others, longer periods of practice are required. Again, patience is necessary.

repetition and habit formation

Skill is characterized by the ability to perform a particular movement with ease and efficiency. Automatic, rapid, accurate, and smooth are other terms that describe skilled performance. As indicated, such skill is developed only through conscious repetition of correct movement patterns. Through such repetition, the nervous system ultimately becomes "grooved" and, given the proper stimulus, the act can be performed without any conscious attention. Much of the so-called "fast thinking" in sports is not deliberate thinking at all. It is a habitual reaction to a situation that a performer has faced many times before. Though he must concentrate on the game, he is still reacting to a particular situation largely by habit. Similarly, the pianist doesn't "think" about where he will place his fingers as he plays. Neither does a typist. Each is depending upon a complex pattern of habitual responses carefully developed by means of endless repetitions—practice.

steps in learning a motor skill

The first step in habit formation leading to skilled performance is an understanding of the pattern to be followed. The learner must first establish in his mind the "general picture" of what he is to do. This is usually accomplished by observing a demonstration of the skill to be performed (either whole or in part, depending on the nature and complexity of the skill), accompanied by a brief, general, verbal explanation of the essential parts of the movement pattern. A film, diagram, or other written explanation may also be helpful, but care must be taken at this stage not to become overwhelmed with details. At the beginning, concentration should be only on getting the "big picture." As a rule, detailed analysis is effective only in the later stages of skill development and at the intermediate or advanced levels.

The next step is an actual trial experience in the movement demonstrated or described. Here the attempt is to imitate the basic movement pattern, again not being overly concerned with the exact copying of every minute detail. Details are generally worked out individually at a later time. Once again, only the general pattern and essential features of the movement should be concentrated upon.

The initial trial is essentially a diagnostic one. It is the instructor's job to point out and reinforce by encouragement the correct movements and to correct improper execution. A repeated demonstration and verbal explanation may again be necessary. At this point, it is important that the learner pay close attention and be actually involved in the analysis of his movement. He should not feel offended or discouraged at any corrections directed towards him. If the essence of a correction is not understood, questions should be asked. A good technique for testing grasp of the movement concept and essential points, as well as for firmly implanting it in your mind, is to repeat the explanation and directions to yourself as you perform the movement.

Following the initial trial, diagnosis, and corrections, repeated practice accompanied by continued instructor evaluation and self-analysis is necessary until an effective movement pattern is established. At first it will be necessary to think through and act out each step carefully and to put the parts together slowly. There will likely be many errors at first. Gradually the pieces of the total performance will fall together and with a sufficient number of repetitions, a smooth performance will result. Thereafter, only the proper stimulus will be needed to set the habit pattern in motion, and with little conscious attention a uniform pattern will be achieved. The habit will then be "grooved."

methods to make practice effective

Practice must be correct. Practice alone doesn't guarantee improvement. It only provides an opportunity. To be effective, practice must first of all be done correctly. Continued repetition of an incorrect technique will only tend to fix an undesirable movement pattern. It is important, then, to have someone watch and correct you in order to avoid forming incorrect habits which will be hard to break.

Importance of self-analysis. Since it is not always possible to have someone watch you, the importance of a clear understanding of the essential elements of correct form and of constant self-analysis cannot be overstressed. In most activities, if the end result of a particular skill execution is unsatisfactory, it can always be analyzed and traced back to a specific cause or causes. For example, if in target archery an archer is consistently hitting high in the center of the target, the cause may be traced to one or a combination of several causes. Among the causes may be opening the mouth just before or during the arrow release, leaning back on or before the release, or jerking the hand back on the release. By being able to trace the error back to its cause, self-correction can be made. Being able to analyze end results and trace errors back to the cause, and knowing the essential elements of proper execution can help you be your own teacher.

Length and spacing of practice sessions. The length and spacing of practice sessions depends on several factors: the complexity of the skill involved, the amount of energy expenditure required, the degree of motivation, and whether the activity is being learned or performance is at an advanced level. In general, short and frequent practice bouts are more effective when first learning an activity if the skills are complex, require a lot of energy, or hold low motivation. Longer practice bouts are generally reserved for skills which are easy to master, enjoyable, and highly motivating, and when the individual skill level is high.

If the practice session (whether complete or a segment) is conducted to the point that boredom or fatigue sets it, the session is too long. Both boredom and fatigue cause a loss of concentration and can result in just "going through the motions" without any real value coming from the practice. Execution becomes careless and bad habits develop. On the other hand, the session must be long enough to accomplish objectives, provide for a proper warm-up, and allow both whole and part practice where appropriate. The sessions must also be frequent enough to allow retention and reinforcement of what has been learned the previous session. The learner should be able to reach at least the level of the previous practice session to overcome in-between forgetting. Daily practice sessions are most effective in establishing the basic skills of an activity. Once established, increasing the intervals between practices gives a better chance for weaknesses to show so that they may be corrected.

Because of cost and scheduling difficulties, practice sessions are often not as frequent as desirable in a school, public, or private instructional program. This makes it necessary for the learner to engage in as much out-of-class practice as possible. Preferably practice should be done with another person. Working with someone else can provide opportunity for mutual observation and correction, and will also make the practice more enjoyable. It is also well established empirically that practicing with someone slightly better than you tends to bring about quicker improvement.

Mental practice. Though physical practice is best, the next best method of improving a skill is through mental practice. Thinking through the execution of a skill helps to establish the motor concept, to develop the neural patterns which control the required muscle movement, and to develop the kinesthetic feel of the skill execution. Many studies have proven the effectiveness of pure mental practice. After being given the same initial demonstration, explanation, and preliminary trials, groups have been divided into those which then engaged in regular physical practice, mental practice, and no practice. Upon retest, though groups performing physical practice were generally superior, remarkable improvements of from 30 to 90 percent were found as a result of mental practice alone, compared with little or no improvement among those who engaged in no type of practice. It appears that a combination of physical and mental practice would be even more effective. Mental practice simply consists of visualizing yourself properly executing the skill repeatedly for five to ten minutes a day. Try it and see for yourself how effective it can be.

Execution speed. Except for the purpose of a keener analysis and understanding of the movement pattern, when attempting to learn or improve a skill, the skill should be practiced at the same speed at which it is to be performed in a game situation. Practicing at a slower speed will develop responses that are only effective at that particular speed. The transfer value to an actual game situation will not be as effective. Often, when accuracy is involved in a particular skill (tennis, golf), there may be a tendency to slow down the movement and even modify the basic elements of good form in an effort to concentrate on placement. This should be avoided. If accuracy is important to the skill, it should be emphasized from the start, but not at the expense of correct form or normal speed.

Proper form. There is really no "best" form for the execution of a particular skill, but there is such a thing as "bad" and "good" form. The basic elements of good form as demonstrated or explained should always be followed during practice, but the fine details are a matter of individual adjustment. For example, essential to the effective stroking of a forehand drive in tennis is having an early backswing, getting the racket in position to thrust the ball at the proper time, stroking with the weight shifted to the front foot, and keeping a firm wrist. How the racket is brought back—whether circular or flat—is relatively unimportant. The exact footwork used to get in position may also vary, as will the path and extent of the follow-through. But failure to have the racket in proper position and slapping at the ball would be serious violations of good form and result in poor execution.

Kinesthetic "feel." Getting the "feel" of the correct action has already been mentioned several times. One other point concerning it in relation to practice should be mentioned. Though you may conscientiously follow all the suggestions about mental practice, analysis, and repetitive physical practice, it may take some time before you actually experience the "feel." When it comes, however, it is unmistakable. Being able to capture the feeling when it occurs and not let it slip by is important. When you have a good execution, try to remember how it felt—stop for awhile and think about it. Get it firmly fixed in your mind. If it happens near the end of a practice period, it is often a good idea to "quit on it." Once captured in this manner, the correct movement is more easily recognized and repeated.

evaluating your progress

In learning various activities, individuals will progress at varying speeds. Of first importance, then, is not to evaluate yourself by comparison with others. Secondly, don't attempt to evaluate your progress by the rate at which you are progressing. Instead, evaluate yourself only in terms of the quality of your performance, no matter how slow your progression. If you make a conscientious effort and avoid becoming anxious or discouraged, improvement is inevitable.

Evaluating yourself, of course, is difficult. Many times, especially if you have a tendency to strive for perfection and are hypercritical of yourself, you may be completely unaware of your improvement. Someone else is often in a better position to make the judgment—certainly your instructor. Since it is an important motivating factor to know the results of your efforts, do not be afraid to ask your instructor for an evaluation, should he not provide it himself. You should also be aware that improvement (like physical growth) is often difficult to notice on a daily basis or over a short period. Here again patience is important. Thinking back at the end of an instructional program to how you were at the start is often the clearest way of noticing improvement.

In some activities, improvement can be measured by the scores made or by comparative results of competitive play. During the early stages of learning, however, care should be taken not to place undue emphasis on the results of competitive play. Overemphasis on scores or competitive results can cause you to neglect proper form and be more "successful" at lower levels of competition, but chances are that you will never rise to a much higher level. The person who concentrates on proper execution, even if it means scoring poorly or "losing" for awhile, will eventually reach a much higher level of skill and enjoyment.

learning curves and retrogression

To avoid undue discouragement, it should be recognized that improvement in physical activities rarely proceeds smoothly and at a uniform rate. Normally, a rapid period of improvement is made early and is followed by a period of slow progress, a plateau, or even retrogression before a noticeable improvement is made again. As in training to raise your level of physical fitness, as you near your maximum capacities improvement will become more difficult.

The causes of this unsteady pattern of improvement are many: lapses in motivation, boredom, emotional difficulties, fatigue, changes in overall level of condition, physical and psychological limits, and adjustment to new skills, particularly when it involves breaking old habits.

When boredom, fatigue, and emotional difficulties combine to thwart motivation and result in a prolonged plateau or period of retrogression, it is sometimes helpful to temporarily break the practice pattern or to even get away completely from the activity for awhile. The temporary change in routine will help to create a new perspective upon resuming the activity. A drop in overall level of physical condition—either from illness or neglect—will always adversely affect performance.

A common cause of retrogression is the adjustment to new skills, especially when it involves breaking or "unlearning" old habits. Many persons sometimes enter an activity having had some previous experience, but without the benefit of sound instruction. They may also have had previous instruction in a particular technique which may not be as effective as another in terms of potential for greater improvement. Changing over to a new technique will obviously require some time before effective results are seen. Having had some limited success with the "old way" and developed a feeling of comfortableness with it, resistance to change is natural. The tendency to revert to the old habit is magnified when the early results of the new technique are seen as worse than when the learner did it "his way." In bowling, for example, the "hook ball" is both theoretically and experimentally a more effective delivery than a "straight ball." But in making the change, a person will feel awkward and will normally score much lower until he becomes adjusted to the new method of release. Many will become discouraged at this and go back to the "old way." By failing to recognize and accept the retrogression as normal and temporary, they limit the ultimate skill level they can reach. The same is true in many other activities where a person may have achieved a moderate success using basically inferior techniques.

Another common cause of early retrogression is the normal physiological adjustment of the body to the specific physical demands of the activity. Even if you are at a good level of condition, repeated specific and unaccustomed movements in a certain activity will cause some muscle soreness and stiffness. This will naturally affect performance adversely for awhile, but as the adjustment is made and the soreness disappears, execution will improve.

maximum challenge level

The importance of interest and motivation for effective learning has been emphasized several times. A major factor in maintaining a high intrinsic motivation is establishing performance goals that will provide an optimum level of challenge; that is, a level of challenge that will stimulate your best efforts. The goals set must be based on a realistic evaluation of your present level of skill and not be either too high or too low. Setting your goals too high, such as attempting to "break par" on your first round of golf will only result in frustration and discouragement. Having no goal or setting it too low on the other hand would be equally ineffective. You must have something to shoot for that will be neither too hard to achieve (frustration) nor too easy (boredom). Using golf as an example, an early goal might be attempting to make a triple bogie (3 strokes over par) on each hole. Once achieved, or if it proves too easy, a higher goal can be set (double bogie: 2 strokes over par). The same procedure can be established for other activities. In an activity such as tennis or handball, a self-made goal may be attempting to execute a particular shot with a certain percentage of accuracy. Depending on comparative levels of ability, a goal in interpersonal competition might be to score a certain number of points against your opponent or to limit him to a certain number. By setting and achieving goals and subgoals, motivation can be constantly stimulated and reinforced, leading to greater enjoyment and improvement.

retention of skill

Once physical skills are learned, unlike verbal learning, they are easily retained no matter how long an absence from participation there might be. Though your level of skill may be lower for a time or even permanently, because of the normal physiological losses of age, the basic ability to perform the skill remains. The explanation is the greater number of stimuli involved in establishing motor concepts and neural patterns than those involved when learning a verbal concept. In addition to the stimuli of sight and sound, there are thousands of additional stimuli sent to the brain while engaging in patterns of movement. Because of this greater neural activity, motor concepts or patterns are more firmly established and are easier to retain than verbal material.

It is never too late to learn physical skills, but the earlier it is done, the greater likelihood of success and continuance. Studies have shown that most adults tend to continue in activities learned in their early twenties or before. At later ages, a major reason for not attempting to learn an activity is social embarassment from the fear of "making a fool of oneself" and appearing inadequate. If it can be remembered that these fears are self-made and that most people admire the learning efforts of older persons, these fears need not be deterrents. Taking advantage of opportunities early in life, on the other hand, can provide a longer lifetime of benefits.

summary

In this book the emphasis has been on presenting a broader concept of fitness for today's world and on developing a personalized, lifetime fitness program based on a solid understanding of the multiple facets of fitness presented and on your specific needs, interests, and lifestyle. The fitness evaluation gave you a more accurate insight into your physical capacities and needs and into methods for evaluating your progress. Exercise principles, methods, precautions, and a variety of specific individual exercise programs were presented to help you safely and effectively meet your specific needs. The importance of sound practices relating to smoking, drinking, drugs, rest, relaxation, and recreation have been emphasized as integral parts of your overall fitness program—without which the benefits gained through exercise would be lessened.

The aim was to help you reach and maintain an adequate level of fitness throughout life in order to offset the mounting health problems associated with our present-day patterns of living and the ignorance or neglect of basic habits of fitness. Hopefully, it will lead you not only to a longer life, but to a life that is lived to its fullest potential and enjoyment.

appendix
a

This index gives you an estimate of your chance at suffering a cardio-vascular accident (heart attack or stroke). An increase in any of the eight risk factors pushes you closer to heart trouble. Check the blank next to the number in each row that best describes you and add up your score. (Row 6: Average U.S. diet is 33% animal or solid fat. Row 7: If you passed insurance exams your blood pressure is probably 140 or less.)

Score Totals

Group 1   6 to 11 = very low risk
Group II   12 to 17 = low risk
Group III   18 to 25 = average risk
Group IV   26 to 32 = high risk
Group V   33 to 42 = dangerous risk
Group VI   42 to 60 = extremely dangerous risk

If your score places you in Groups IV through VI, see your physician. If your score places you in Group III, improvement can be made.

[1] Used through the courtesy of John L. Boyer, M.D., and Director of the San Diego State College Exercise Physiology Laboratory, San Diego, California.

table 11
cardiac risk index

	10 to 20	21 to 30	31 to 40	41 to 50	51 to 60	61 to 70 and over
1. *Age*	1	2	3	4	6	8
2. *Heredity*	No known history of heart disease — 1	1 relative with cardiovascular disease over 60 — 2	2 relatives with cardiovascular disease over 60 — 3	1 relative with cardiovascular disease under 60 — 4	2 relatives with cardiovascular disease under 60 — 6	3 relatives with cardiovascular disease under 60. — 8
3. *Weight*	More than 5 lbs. below standard weight — 0	Standard weight — 1	5–20 lbs. overweight — 2	21–35 lbs. overweight — 3	36–50 lbs. overweight — 5	51–65 lbs. overweight — 7
4. *Tobacco Smoking*	Nonuser — 0	Cigar and/or pipe — 1	10 cigarettes or less a day — 2	20 cigarettes a day — 3	30 cigarettes a day — 5	40 cigarettes a day or more — 8
5. *Exercise*	Intensive occupational and recreational exertion — 1	Moderate occupational and recreational exertion — 2	Sedentary work and intense recreational exertion — 3	Sedentary occupational and moderate recreational exertion — 5	Sedentary work and light recreational exertion — 6	Complete lack of all exercise — 8

6. *Cholesterol or % fat in diet*	Cholesteral below 180 mg. Diet contains no animal or solid fats [1]	Cholesterol 181–205 mg. Diet contains 10% animal or solid fats [2]	Cholesterol 206–230 mg. Diet contains 20% animal or solid fats [3]	Cholesterol 231–255 mg. Diet contains 30% animal or solid fats [4]	Cholesterol 256–280 mg. Diet contains 40% animal or solid fats [5]	Cholesterol 281–330 mg. Diet contains 50% animal or solid fats [7]
7. *Blood Pressure*	100 upper reading [1]	120 upper reading [2]	140 upper reading [3]	160 upper reading [4]	180 upper reading [6]	200 or over upper reading [8]
8. *Sex*	Female [1]	Female over 45 [2]	Male [3]	Bald Male [4]	Bald short male [6]	Bald short stocky male [7]

Total Score: _____

instructions

Check those activities in which you consider yourself competent. To consider yourself competent in beginning swimming you should be able to jump into the deep end of a pool, tread water or float for ten minutes, and then swim to the shallow end and back (approximately 50 yards), using any stroke. For the other activities, consider yourself competent at the beginning level if you would accept an invitation to play or engage in the activity without undue hesitation. If you would feel embarrassed because of poor skill or lack of knowledge (rules, scoring, etc.), you should not consider yourself competent.

I. *Team Sports*
   A. Indoor
      1. Basketball _____
      2. Volleyball _____
      3. _____ _____
   B. Outdoor
      1. Softball _____
      2. Tag Football _____
      3. Soccer _____
      4. _____ _____

II. *Individual-Dual Sports (competitive)*
   A. Indoor
      1. Badminton _____
      2. Bowling _____
      3. Handball _____
      4. Squash _____
      5. _____ _____
   B. Outdoor
      1. Archery _____
      2. Golf _____
      3. Tennis _____
      4. _____ _____

III. *Individual-Group (Recreational)*
   A. Indoor
      1. Bridge _____
      2. Billiards _____
      3. Pool _____
      4. Table Tennis _____

      5. Chess _____
      6. _____ _____
   B. Outdoor
      1. Camping _____
      2. Fishing _____
      3. Riding _____
      4. _____ _____

IV. *Combatives*
      1. Judo _____
      2. Karate _____
      3. Wrestling _____
      4. Fencing _____
      5. _____ _____

V. *Conditioning and Gymnastics*
      1. Individual exercise programs _____
      2. Weight training _____
      3. Gymnastics-apparatus _____
      4. Gymnastics-trampolining _____
      5. Gymnastics-tumbling _____
      6. _____ _____

VI. *Dance*
      1. Social Dance _____
      2. Folk-square dance _____
      3. Modern dance _____
      4. _____ _____

VII. *Aquatics*

1. Beginning Swimming	_____	5. Skin Diving	_____
2. Intermed. Swimming	_____	6. Scuba Diving	_____
3. Lifesaving	_____	7. Surfing	_____
4. Water Safety Instr.	_____	8. Water Skiing	_____
		9. Boating	_____
		10. Sailing	_____
		11. _____	_____

note

Ideally, you should have at least beginning level competency in at least one activity in each major area and subarea (indoor and outdoor). Because of their limited carry-over value, however, team sports and combatives may be excluded.

your ability to relax[2]

	Always	Sometimes	Seldom	Never
1. Do you plan your life to include change, i.e., people, activities, places to go, etc?	___	___	___	___
2. Are you free from worries and moods?	___	___	___	___
3. Do you find time to relax and/or stretch during the day?	___	___	___	___
4. Are you free from nervousness and jittery feelings?	___	___	___	___
5. Are you free from headaches and twitches?	___	___	___	___
6. Do you find yourself scowling, clenching fists, tightening your jaws, hunching your shoulders, or pursing your lips?	___	___	___	___
7. When you find yourself becoming tense because of sustained positions, do you relax by doing simple movements?	___	___	___	___
8. Can you relax evidences of tension at will when you find them?	___	___	___	___
9. If a nap is taken during the day, do you wake up refreshed?	___	___	___	___
10. Are you able to concentrate on one problem at a time?	___	___	___	___
11. In sports, games, or hobbies, do you participate with such interest that you are completely absorbed in what you are doing?	___	___	___	___

[2] From Wayne D. Van Huss, Roy K. Niemeyer, Herbert A. Olson, and John A. Friedrich, *Physical Activity in Modern Living*, 2nd ed. © 1969. Adapted by permission of Prentice-Hall, Inc., Englewood Cliffs, N.J.

12. Do you wake up in the morn-
ing refreshed?   ____   ____   ____   ____

13. Do you find it easy to get to
sleep at night?   ____   ____   ____   ____

14. Are you able to shut out your
worries when you go to bed
at night?   ____   ____   ____   ____

15. Are you able to release tensions through simple movements so you can get to
sleep?   ____   ____   ____   ____

Scoring: Always: 3 points   Ratings: 38–45: High
           Sometimes: 2 points   30–37: Average
           Seldom: 1 point   Under 30: Low
           Never: 0 points

POSTURE SCORE SHEET

SCORING DATES

Name _____

	GOOD – 10	FAIR – 5	POOR – 0			
HEAD LEFT RIGHT	HEAD ERECT GRAVITY LINE PASSES DIRECTLY THROUGH CENTER	HEAD TWISTED OR TURNED TO ONE SIDE SLIGHTLY	HEAD TWISTED OR TURNED TO ONE SIDE MARKEDLY			
SHOULDERS LEFT RIGHT	SHOULDERS LEVEL (HORIZONTALLY)	ONE SHOULDER SLIGHTLY HIGHER THAN OTHER	ONE SHOULDER MARKEDLY HIGHER THAN OTHER			
SPINE LEFT RIGHT	SPINE STRAIGHT	SPINE SLIGHTLY CURVED LATERALLY	SPINE MARKEDLY CURVED LATERALLY			
HIPS LEFT RIGHT	HIPS LEVEL (HORIZONTALLY)	ONE HIP SLIGHTLY HIGHER	ONE HIP MARKEDLY HIGHER			
ANKLES	FEET POINTED STRAIGHT AHEAD	FEET POINTED OUT	FEET POINTED OUT MARKEDLY ANKLES SAG IN (PRONATION)			
NECK	NECK ERECT, CHIN IN, HEAD IN BALANCE DIRECTLY ABOVE SHOULDERS	NECK SLIGHTLY FORWARD, CHIN SLIGHTLY OUT	NECK MARKEDLY FORWARD, CHIN MARKEDLY OUT			
UPPER BACK	UPPER BACK NORMALLY ROUNDED	UPPER BACK SLIGHTLY MORE ROUNDED	UPPER BACK MARKEDLY ROUNDED			
TRUNK	TRUNK ERECT	TRUNK INCLINED TO REAR SLIGHTLY	TRUNK INCLINED TO REAR MARKEDLY			
ABDOMEN	ABDOMEN FLAT	ABDOMEN PROTRUDING	ABDOMEN PROTRUDING AND SAGGING			
LOWER BACK	LOWER BACK NORMALLY CURVED	LOWER BACK SLIGHTLY HOLLOW	LOWER BACK MARKEDLY HOLLOW			
REEDCO INCORPORATED			**TOTAL SCORES**			

*Reproduced through the courtesy of REEDCO, Inc., Auburn, New York.
(Score Sheet pads and related posture analysis materials available through REEDCO, Auburn, New York.)

Fig. 12. Posture Analysis Form

**general health
and fitness check list**[3]

	Yes	*No*
1. Have you had a medical exam during the past year?	____	____
2. Have you had a chest X-ray during the past year?	____	____
3. Were the recommendations of your doctor followed, if any?	____	____
4. Have you visited the dentist during the last year?	____	____
5. Has all indicated dental work been completed?	____	____
6. Have you followed the recommendations he gave you?	____	____
7. Have you had your eyes tested within the past five years?	____	____
8. Are you within the limit for obesity on the basis of the body fat measurements taken and your particular body type?	____	____
9. Are you free from postural defects?	____	____
10. Are you free from foot trouble, joint pains, or backache?	____	____
11. Was the result of your 12-minute test at least good?	____	____
12. Was your Heart Recovery Rate at least good?	____	____
13. Were you within normal limits on the vital capacity tests?	____	____
14. Are you a nonsmoker, or, if so do you smoke less than a half pack per day?	____	____
15. Do you avoid the frequent use of aspirin, laxatives, tranquilizers, or energizers?	____	____
16. Do you eat a balanced diet, including a good breakfast, lunch and dinner?	____	____
17. Do you avoid using too much salt in your food or eating too many high calorie and saturated fat foods?	____	____
18. Do you have at least beginning level skills and knowledge in at least 3 indoor and 3 outdoor activities other than major team sports, including swimming?	____	____
19. Do you have at least one or two hobbies that you enjoy?	____	____

[3] From Wayne D. Van Huss, Roy K. Niemeyer, Herbert A. Olson, and John A. Friedrich, *Physical Activity in Modern Living*, 2nd ed. Copyright © 1969. Adapted by permission of Prentice-Hall, Inc., Englewood Cliffs, N.J.

Yes   *No*

20. Are you free from undue worry and able to relax when you should?

21. Do you sleep enough and well enough so that you feel rested when you get up?

22. Are you seldom drowsy during the day?

23. Are you able to get through the day without feeling unduly fatigued?

Any negative responses would indicate a need to consider possible corrective measures.

fitness evaluation summary

Name _____ Sex _____ Age _____

  I. *Medical Examination*
     Limitations:_____

_____

 II. *Body Type*
     General Classification:_____

III. *Posture Analysis*
     Deviations Noted:_____

_____

IV. *Body Proportions and Body Fat Analysis*

Body Proportions					*Fat Analysis*				
Date	___	___	___	___	*Pinch Test* (Amount over for Body				
Height	___	___	___	___	Type):				
Weight	___	___	___	___	Ecto (1/2″)	___	___	___	___
Chest	___	___	___	___	Meso (3/4″)	___	___	___	___
Waist	___	___	___	___	Endo (1″)	___	___	___	___
Hips	___	___	___	___	*Skinfold Caliper Measurements:*				
Thigh	___	___	___	___	Bicep	___	___	___	___
Calf	___	___	___	___	Tricep	___	___	___	___
					Hip	___	___	___	___
					Subscapular	___	___	___	___
					Total	___	___	___	___
					Body Fat %	___	___	___	___

  V. *Cardiorespiratory Fitness*

A. Resting Heart Rate
Lying, A.M. (1 min.)   ___ ___
Sitting, after 5 min. lying
  1 min.   ___ ___
  10 sec.   ___ ___
B. Heart Recovery Test
(10 sec. counts, standing)
Starting HR   ___ ___
Immediately after exercise   ___ ___
1 min. after exercise   ___ ___

C. 12 Minute Run-Walk
Distance   ___ ___
Rating   ___ ___
D. Vital Air Capacity
$PFEV_{1.0}$ ___ ___
Observed   ___ ___
PFVC   ___ ___
Observed   ___ ___
E. Cardiac Risk Index
Score   ___ ___
Risk   ___ ___

VI. *Activity Competency*
     Areas Needing Competency:_____

_____

VII. *Ability to Relax* (Check One.)
　　　High_____ 　Average_____ 　Low_____
VIII. *General Health and Fitness*
　　　Number of Negative Responses:_____
　　　Corrective Measures Indicated:_____

　　　_____
　　　_____
　　　_____

appendix
b

Four short tests follow to help you find out what you *know* about cigarette smoking and how you *feel* about it. They can tell you:

1. Whether you really want to quit smoking
2. What you know about the effects of smoking on health
3. What kind of smoker you are (why you smoke)
4. Whether the world you live in will help or hinder you if you do try to stop

We believe that if you take a good hard look at the facts and that if you analyze your real feelings, you may decide to quit smoking. Tests 1 and 2 are designed to help you take this look at yourself. Tests 3 and 4 will give you some insight into what kind of smoker you are, and will reveal some of the problems you may run into when you try to quit.

[1] From U.S. Department of Health, Education, and Welfare Public Service Publication No. 2013, *The Smoker's Self-Test* (Washington, D.C.: U.S. Government Printing Office, December, 1969).

test 1
do you want to change
your smoking habits?

For each statement, circle the number that most accurately indicates how you feel. For example, if you completely agree with the statement, circle 4; if you agree somewhat, circle 3, etc. *Important: Answer every question.*

	Completely Agree	Somewhat Agree	Somewhat Disagree	Completely Disagree
A. Cigarette smoking might give me a serious illness.	4	3	2	1
B. My cigarette smoking sets a bad example for others.	4	3	2	1
C. I find cigarette smoking to be a messy kind of habit.	4	3	2	1
D. Controlling my cigarette smoking is a challenge to me.	4	3	2	1
E. Smoking causes shortness of breath.	4	3	2	1
F. If I quit smoking cigarettes it might influence others to stop.	4	3	2	1
G. Cigarettes cause damage to clothing and other personal property.	4	3	2	1
H. Quitting smoking would show that I have willpower.	4	3	2	1
I. My cigarette smoking will have a harmful effect on my health.	4	3	2	1
J. My cigarette smoking influences others close to me to take up or continue smoking.	4	3	2	1
K. If I quit smoking, my sense of taste or smell would improve.	4	3	2	1
L. I do not like the idea of feeling dependent on smoking.	4	3	2	1

how to score

1. Enter the numbers you have circled to the Test 1 questions in the spaces below, putting the number you have circled to Question A over line A, to Question B over line B, etc.

2. Total the 3 scores across on each line to get your totals. For example, the sum of your scores over lines A, E, and I gives you your score on *Health*—lines B, F, and J give the score on *Example*, etc.

Totals

_____ + _____ + _____ = _____
  A   **E**   **I**   Health

_____ + _____ + _____ = _____
  B   **F**   **J**   Example

_____ + _____ + _____ = _____
  C   **G**   **K**   Esthetics

_____ + _____ + _____ = _____
  D   **H**   **L**   Mastery

Scores can vary from 3 to 12. Any score 9 and above is *high;* any score 6 and below is *low.*

test 2

what do you think
the effects of smoking are?

For each statement, circle the number that shows how you feel about it. Do you strongly agree, mildly agree, mildly disagree, or strongly disagree? *Important: Answer every question.*

	Strongly Agree	Mildly Agree	Mildly Disagree	Strongly Disagree
A. Cigarette smoking is not nearly as dangerous as many other health hazards.	1	2	3	4
B. I don't smoke enough to get any of the diseases that cigarette smoking is supposed to cause.	1	2	3	4
C. If a person has already smoked for many years, it probably won't do him much good to stop.	1	2	3	4
D. It would be hard for me to give up smoking cigarettes.	1	2	3	4
E. Cigarette smoking is enough of a health hazard for something to be done about it.	4	3	2	1
F. The kind of cigarette I smoke is much less likely than other kinds to give me any of the diseases that smoking is supposed to cause.	1	2	3	4
G. As soon as a person quits smoking cigarettes he begins to recover from much of the damage that smoking has caused.	4	3	2	1
H. It would be hard for me to cut down to half the number of cigarettes I now smoke.	1	2	3	4

I. The whole problem of cigarette smoking and health is a very minor one.	1	2	3	4
J. I haven't smoked long enough to worry about the diseases that cigarette smoking is supposed to cause.	1	2	3	4
K. Quitting smoking helps a person to live longer.	4	3	2	1
L. It would be difficult for me to make any substantial change in my smoking habits.	1	2	3	4

how to score

1. Enter the numbers you have circled to the Test 2 questions in the spaces below, putting the number you have circled to Question A over line A, to Question B over line B, etc.

2. Total the 3 scores across on each line to get your totals. For example, the sum of your scores over lines A, E, and I gives you your score on *Importance*—lines B, F, and J give the score on *Personal Relevance*, etc.

Totals

_____ + _____ + _____ = _____
 A   **E**   **I**   Importance

_____ + _____ + _____ = _____
 B   **F**   **J**   Personal Relevance

_____ + _____ + _____ = _____
 C   **G**   **K**   Value of Stopping

_____ + _____ + _____ = _____
 D   **H**   **L**   Capability for Stopping

Scores can vary from 3 to 12. Any score 9 and above is *high;* any score 6 and below is *low.*

test 3

why do you smoke?

Here are some statements made by people to describe what they get
out of smoking cigarettes. How *often* do you feel this way when smoking
them? *Important: Answer every question.*

		Always	Fre-quently	Occa-sionally	Seldom	Never
A.	I smoke cigarettes in order to keep myself from slowing down.	5	4	3	2	1
B.	Handling a cigarette is part of the enjoyment of smoking it.	5	4	3	2	1
C.	Smoking cigarettes is pleasant and relaxing.	5	4	3	2	1
D.	I light up a cigarette when I feel angry about something.	5	4	3	2	1
E.	When I have run out of cigarettes I find it almost unbearable until I can get them.	5	4	3	2	1
F.	I smoke cigarettes auto-matically without even being aware of it.	5	4	3	2	1
G.	I smoke cigarettes to stimu-late me, to perk myself up.	5	4	3	2	1
H.	Part of the enjoyment of smoking a cigarette comes from the steps I take to light up.	5	4	3	2	1
I.	I find cigarettes pleasurable.	5	4	3	2	1
J.	When I feel uncomfortable or upset about something, I light up a cigarette.	5	4	3	2	1
K.	I am very much aware of the fact when I am not smoking a cigarette.	5	4	3	2	1
L.	I light up a cigarette without realizing I still have one burning in the ashtray.	5	4	3	2	1

M. I smoke cigarettes to give me a "lift."	5	4	3	2	1
N. When I smoke a cigarette, part of the enjoyment is watching the smoke as I exhale it.	5	4	3	2	1
O. I want a cigarette most when I am comfortable and relaxed.	5	4	3	2	1
P. When I feel "blue" or want to take my mind off cares and worries, I smoke cigarettes.	5	4	3	2	1
Q. I get a real gnawing hunger for a cigarette when I haven't smoked for a while.	5	4	3	2	1
R. I've found a cigarette in my mouth and didn't remember putting it there.	5	4	3	2	1

how to score

1. Enter the numbers you have circled to the Test 3 questions in the spaces below, putting the number you have circled to Question A over line A, to Question B over line B, etc.

2. Total the 3 scores across on each line to get your totals. For example, the sum of your scores over lines A, G, and M gives you your score on *Stimulation*—lines B, H, and N give the score on *Handling*, etc.

Totals

___ +	___ +	___ =	___	
A	**G**	**M**	Stimulation	
___ +	___ +	___ =	___	
B	**H**	**N**	Handling	
___ +	___ +	___ =	___	
C	**I**	**O**	Pleasurable Relaxation	
___ +	___ +	___ =	___	
D	**J**	**P**	Crutch: Tension Reduction	
___ +	___ +	___ =	___	
E	**K**	**Q**	Craving: Psychological Addiction	
___ +	___ +	___ =	___	
F	**L**	**R**	Habit	

Scores can vary from 3 to 15. Any score 11 and above is *high;* any score 7 and below is *low.*

test 4

does the world around you make it easier or
harder to change your smoking habits?

Indicate by circling the appropriate numbers whether you feel the
following statements are true or false. *Important: Answer every question.*

	True or *Mostly True*	*False or* *Mostly False*
A. Doctors have decreased or stopped their smoking of cigarettes in the past 10 years.	2	1
B. In recent years there seem to be more rules about where you are allowed to smoke.	2	1
C. Cigarette advertising makes smoking appear attractive to me.	1	2
D. Schools are trying to discourage children from smoking.	2	1
E. Doctors are trying to get their patients to stop smoking.	2	1
F. Someone has recently tried to persuade me to cut down or quit smoking cigarettes.	2	1
G. The constant repetition of cigarette advertising makes it hard for me to quit smoking.	1	2
H. Both Government and private health organizations are actively trying to discourage people from smoking.	2	1
I. A doctor has, at least once, talked to me about my smoking.	2	1
J. It seems as though an increasing number of people object to having someone smoke near them.	2	1
K. Some cigarette commercials on TV make me feel like smoking.	1	2
L. Congressmen and other legislators are showing concern about smoking and health.	2	1

M. The people around you, particularly those who are close to you (e.g., relatives, friends, office associates), may make it easier or more difficult for you to give up smoking by what they say or do. What about these people? Would you say that they make giving up smoking or staying off cigarettes more difficult for you than it would be otherwise? (Circle the number to the left of the statement that best describes your situation.)
3 They make it much more difficult than it would be otherwise.

test 4 (cont.)

4 They make it somewhat more difficult than it would be otherwise.
5 They make it somewhat easier than it would be otherwise.
6 They make it much easier than it would be otherwise.

how to score

1. Enter the numbers you have circled to the Test 4 questions in the spaces below, putting the number you have circled to Question A over line A, to Question B over line B, etc.

2. Total the 3 scores across on each line to get your totals. For example, the sum of your scores over lines A, E, and I gives you your score on *Doctors*—lines B, F, and J give the score on *General Climate*, etc.

Totals

___ + ___ + ___ = _____
 A   E   I   Doctors

___ + ___ + ___ = _____
 B   F   J   General Climate

___ + ___ + ___ = _____
 C   G   K   Advertising Influence

___ + ___ + ___ = _____
 D   H   L   Key Group Influences

         ___ = _____
          M   Interpersonal Influences

Scores can vary from 3 to 6: 6 is *high;* 5, high middle; 4, low middle; 3, *low.*

breaking the habit

The difficulty one experiences in attempting to quit smoking and whether or not one is successful depends partly on the reasons he smokes, the environmental influences about him, his knowledge and convictions about the possible damage it is causing, and the type and strength of the motivation to stop. Many people, though convinced of the evidence against smoking, suffer from the common "immunity complex" all humans seem to have—the belief that they will be the one who escapes the consequences or is the exception to the statistics. It's something that happens to others, but not to them. For them, only the "immediate threat" to health is an effective health motivation. More common and effective motivations are ego (the desire to gain mastery over a habit that has gained control over a person's behavior), being shamed by family or children and the desire to set an example, and offense to individual esthetic sense (unpleasant taste or odor, messiness, damage to clothing and furniture, etc.). For young people, the desire to achieve top condition for athletic participation is one of the most effective motivations.

No method or gimmick for quitting can work without some kind of motivation and a real desire to stop, but the reason for smoking will determine the degree of difficulty encountered and possibly the best method. Those who smoke out of craving or psychological addiction or for whom cigarette smoking is a crutch in tense situations are referred to as "negative smokers." These types usually have the most difficulty in quitting unless substitutes can be found for satisfying their cravings and handling their tensions. "Positive smokers," or those who smoke for relaxation, enjoyment, self-reward, etc., generally have less difficulty. Though there is often a combination of elements in any one individual smoker, most can be generally classified into one type or the other.

Once motivation is established, a decision made, and some insight gained into the reasons for smoking, the basis is laid for determining the method of breaking the habit. Two basic methods are a "cold turkey" abrupt withdrawal or a "gradual withdrawal," supplemented by a variety of techniques aimed at gaining greater insight into smoking behavior patterns, altering behavior, and reinforcing new habit patterns. Essential is the avoidance of any feeling of being deprived, the development of a new routine for handling "real" needs formerly met by cigarettes, and a positive attitude about the ability to succeed.

Some medical personnel feel that for "negative smokers," a "cold turkey" withdrawal, with plenty of substitutes (gum, candy, etc.), lots of exercise and rest, and the possible use of tranquilizers is the most effective method. "Positive smokers," they feel, can effectively break the habit by gradual withdrawal, but not the negative smokers. According to the American Cancer Society, however, five out of six persons who quit cold turkey later resumed. The success experienced by the private "Smoke Watchers" organization and other public withdrawal clinics utilizing the gradual withdrawal method appears to further point to this method as being the most effective for all. Psychologically, it is sound. Since the smoking habit is gradually developed, it is logical that the reversal and development of new habits must also be gradual. The feeling of deprivation and the tendency towards resentment is also lessened in this way.

The Smoke Watcher's technique (claimed 97 percent successful for 7,000 smokers) essentially involves filling out a weekly cigarette chart showing the time, activity engaged in, feeling (anger, boredom, joy, etc.), level of desire, and time length of each cigarette smoked. The procedure is aimed at helping the smoker gain a greater awareness of the nature of his smoking habit, to pinpoint the "pure habit" cigarettes (unconscious lighting up) and those that meet specific needs of varying degrees at various times. The smoker quickly comes to eliminate the "habit" cigarettes, to separate the others into "important" and "unimportant" ones, and to gradually eliminate the less needed ones. Merely taking time to jot down the information or waiting for the opportunity to do so is often enough to ward off the desire. After seven or eight weeks, even heavy smokers have reportedly come down to five or six "hard core" cigarettes. In eight to twelve weeks, the desire as well as the need is often eliminated. Significant in this gradual withdrawal approach is the need to take a relatively long-range view of results, rather than expecting immediate changes, in order to avoid discouragement and allow for the gradual development of new behavior patterns. This long-range view is similar to that emphasized for weight control and exercise training programs.

Dr. Donald Friedrickson, a New York City public health official, reported very successful results using a similar technique at a free withdrawal clinic. Smokers were instructed to fill out a similar chart and to assign a value of 1 to 5 for each cigarette based on need and desire. An added gimmick was "pack wrapping": keeping the cigarette package wrapped in a paper or cloth to add a slight inconvenience to the normally easy procedure of reaching for a cigarette. It was so bothersome that many simply "forgot about it." A 75 percent reduction was reported common.

Other successfully used gimmicks during gradual withdrawal include: listing the reasons for quitting on a card and carrying it around with you; substituting gum or suckers; reaffirming the decision to quit after every successful warding off of a desire; buying just one pack at a time and keeping it out of sight or in inconvenient places; changing brands often, each time to one with a lower nicotine and tar content; keeping ashtrays out of sight or filling them with paper clips; not carrying matches; smoking at only certain times and in certain places; putting money normally spent on cigarettes in a jar and buying a special purchase with it; and many more. There is no magic formula, but anyone with a desire and a plan spaced over a period of time to allow for a gradual change can stop.

Combining the effort with a program of exercise, pleasurable recreational activities, and adequate rest can also help. A running program aimed at a specific goal can be especially helpful, since the futility of attempting to reach the goal while continuing to smoke will easily become apparent, as well as the ease and improvement in running as cigarette consumption is lessened. The deep breathing and voluntary physical relaxation techniques presented in Chapter 8 can also be valuable aids, along with the other suggestions for reducing stress and tension.

benefits of quitting

The benefits of breaking the smoking habit are many and real. Even without an exercise program, studies of post-quitting physiological effects showed a decrease of 10 percent in oxygen consumption and a 5 percent decrease in the heart rate. Personal studies of post-quitting effects on persons engaged in exercise programs showed decreases in heart rates of up to 22 percent. One who quits smoking can expect to have a longer life, reduce the risk of disability, eliminate the "cigarette cough," profit economically, have the sense of smell return, and enjoy the return of taste while eliminating the odor and discomforts of the cigarette habit.

The weight gain experienced by many who quit is not necessarily from overeating to compensate for oral gratification lost or to overcome nervousness as many people suspect. Though this can be a factor, experiments have shown that even when food intake remained the same, there was an average gain of 6 1/2 pounds in unexercised subjects as a result of a slower metabolism. Those who followed an exercise program after quitting, however, showed no gain.

For those who can't or won't quit completely, switching to filter, low tar-nicotine, and longer cigarettes and smoking them just beyond halfway can help to lessen damage. A New York study showed a drop in the lung cancer rate among those who switched to filter cigarettes, and a lower relative risk for those who smoked filters at least ten years after the switch. Since the tobacco even in nonfilter cigarettes helps to filter out tars, the tar concentration becomes heavier as the cigarette is smoked and is heaviest during the last inch or so.

evaluation guides for drinkers

Because most evidence of serious physical damage is associated primarily with heavy drinkers, the general consensus of professional opinion is that occasional, light drinking presents limited physiological danger. To the majority of persons who drink, it is associated with a "good time" or is used as a valid method of relaxation. (The same can be said for many who use marijuana, though the relative dangers have not as yet been clearly established and there is some question as to whether or not moderate use is possible in terms of effects.) At any rate, it should be pointed out that one-third of all adults demonstrate by their complete abstinence that alcohol is not necessary for either a good time or for relaxation; even for the social drinker there are some dangers.

A very real danger is a failure to take proper precautions with respect to driving or performing other demanding tasks under the influence of alcohol. Swedish driving tests, for example, have shown skill and perception losses of up to 30 percent from average alcohol concentration levels in the blood of .049 percent. (One mixed drink, two 12 oz. cans of beer, or two 3 1/2 oz. glasses of dry wine equal .03 percent.) Impaired judgement, coordination, reaction, vision, and hearing make driving definitely dangerous at .12 percent concentrations, and at .15 percent effects are compounded and a person is considered legally under the influence of alcohol in most states. Another danger of even occasional or moderate drinking at lower levels of concentration is if a person's behavior is changed in such a way that he loses control of his reason and acts contrary to his normal behavior or in a way that is harmful to himself or to others. The most serious potential danger of all, however, is alcoholism.

what is an alcoholic?

Technically, an alcoholic is a person who has lost control over his drinking. He become a "compulsive drinker," unable to resist the urge to drink or, once started, to stop until intoxicated into oblivion. Practically, it is anyone whose drinking habits have a serious effect on his health or interfere with his normal pattern of activities, his personality, reputation, job, or home responsibilities. Skid row alcoholics account for less than 5 percent of all alcoholism. The remaining 95 percent come from all walks of life in both rural and city communities—doctors, lawyers, clergymen, skilled and unskilled laborers, housewives—of every race, nationality, and economic level.

causes of alcoholism

The specific causes of alcoholism are still not well understood. For some, it is a biological inability to tolerate alcohol, similar to the problem of a diabetic. For most, the causes are multiple: a complex combination of psychological, emotional, and environmental factors which ultimately result in a physical dependence upon alcohol that can be broken only with the most intensive medical, psychiatric, and psychological care and counseling, together with a strong personal effort. Support of this theory, rather than alcohol intolerance per se, comes from case histories which show an easy transference of dependence from alcohol to drugs and the similarity of treatment and recovery.

characteristics of the alcoholic

Typically, the alcoholic has a very low frustration level and tends to overreact emotionally to difficult situations. General behavior is also characterized by irrational conduct; a lack of responsibility and emotional restraint; an uncertain, fearful, or antisocial attitude; a feeling of help-lessness; and a tendency to escape realities.

development

Alcoholism is usually a progressively addictive disease, though some discover an addiction from the first drink (physical dependence, with-drawal symptoms, tolerance buildup). It usually begins, however, by relying on a drink to get over a difficult time (psychological dependence). Ironically, the ability to handle the difficulty is lessened and the problem only becomes worse. A vicious cycle begins, along with a physical dependence, tolerance buildup, and withdrawal symptoms, until the alcohol becomes a basic need as well as a crutch to solve personal, pro-fessional, and social problems. Drink becomes the alcoholic's "best friend," substituting for normal human relations and replacing them as his major concern.

Others begin drinking by seeing it as a harmless, simple pleasure pro-viding a sense of relaxation, a "social lubricant," and a feeling of fellow-ship. They continue as just "social" or "good time" drinkers, feeling able to stop anytime they want to—only to find that they no longer can. This is perhaps the greatest risk facing the social drinker: he can never be certain he doesn't have the physiological and emotional characteristics that can lead to alcoholism.

warning signs

The most important clue to a tendency towards alcoholism is the reason for drinking. If you drink because of frustration or nervousness, to forget a problem or worry, or to "pep you up," then be careful. Feeling that you "should" or "must" have just a few drinks before an important engagement, appointment, or social gathering in order to "feel loose" or "have a good time" is another warning sign. More dangerous signs of approaching alcoholism include:[2]

1. Beginning to worry about the supply of alcohol on hand
2. A tendency to consistently drink more than you intended to and beginning to make promises to yourself and others about controlling your drinking
3. Beginning to excuse yourself or lie about your drinking
4. Becoming antisocial or belligerent with others when drinking
5. Beginning to drink alone, taking "eye openers," and going on benders
6. Developing deep feelings of remorse, resentment, and anxiety after drinking bouts

[2] Adapted with permission of The Macmillan Company from *Health: A Quality of Life* by John S. Sinacore. Copyright © by John S. Sinacore, 1968.

selected references

chapter 1

American Association of Health, Physical Education, and Recreation, *Health Facts.* Washington, D.C., 1967.

American Heart Association, *Reduce Your Risk of Heart Attack.* New York, 1968.

————, *Smoking and Heart Disease.* New York, 1968.

BLAKESLEE, ALTON, *What You Should Know About Drugs and Narcotics.* New York: The Associated Press, 1969.

BLAKESLEE, ALTON, and JEREMIAH STAMLER, M.D., *Your Heart Has Nine Lives.* Englewood Cliffs, N.J.: Prentice-Hall, Inc., 1963.

LAMB, LAWRENCE E., M.D., "Living With Your Heart," a series of six articles appearing in *The San Diego Union,* October 5–10, 1969. Presented by the World Book Science Service.

MAISEL, ALBERT Q., "Alcohol and Your Brain: Some News for Social Drinkers," *Readers Digest,* June, 1970.

MONTOYE, H. J., "Summary of Research on the Relationship of Exercise to Heart Disease," *Journal of Sports Medicine and Physical Fitness,* 2 (1962), 35–43.

National Coordinating Council on Drug Abuse Education, *Grassroots.* Washington, D.C.: Drug Abuse Information Service. Provides a resource list along with new and updated drug abuse information on a monthly basis.

National Tuberculosis and Respiratory Disease Association, *Chronic Obstructive Pulmonary Disease—A Manual for Physicians.* Portland, Oregon: Oregon Thoracic Society, 1966.

"Packing Up a 20-Year Study of Heart Ills," *Medical World News,* 37 (1971), 21.

The San Diego Union, "U.S. Health Record One of Poorest," July 12, 1970, A-1.

U.S. Department of Health, Education, and Welfare, *Chart Book on Smoking, Tobacco, and Health,* Public Health Service Publication No. 1937. Washington, D.C., June, 1969; and *The Facts About Smoking and Health,* Public Health Service Publication No. 1712, rev. October, 1968.

———, *Resource Book for Drug Abuse Information.* Chevy Chase, Maryland: National Institute of Mental Health, October 1969.

chapter 2

COOPER, KENNETH H., M.D., *Aerobics.* New York: M. Evans & Company, Inc., 1968.

———, *The New Aerobics.* New York: M. Evans & Company, Inc., 1970.

DE VRIES, HERBERT A., *Physiology of Exercise.* Dubuque, Iowa: William C. Brown Company, 1966. See especially the chapter, "Metabolism and Weight Control."

DOOLITTLE, T. L., and ROLLIN BIGBEE, "The 12-minute Run-Walk: A Test of Cardiorespiratory Fitness for Adolescent Boys," *Research Quarterly,* 39 (1968), 491–95.

EAMES, DAN H., M.D., *The Annual Medical Examination—What Should It Consist Of?* 1967. Pamphlet published at authors' expense.

FLEISHMAN, EDWIN A., *The Structure and Measurement of Physical Fitness.* Englewood Cliffs, N.J.: Prentice-Hall, Inc., 1964.

HASKINS, MARY JANE, *Evaluation in Physical Education.* Dubuque, Iowa: William C. Brown Company, 1971.

KARPOVICH, PETER V., and WAYNE E. SINNING, *Physiology of Muscular Activity* (7th ed.). Philadelphia: W. B. Saunders Company, 1971.

MCARDLE, WILLIAM D., LINDA ZWIREN, and JOHN R. MAGEL, "Validity of the Post Exercise Heart Rate as a Means of Estimating Heart Rate During Work of Varying Intensities," *Research Quarterly,* 40, No. 3 (1969), 523–28.

MAYER, JEAN, *Overweight.* Englewood Cliffs, N.J.: Prentice-Hall, Inc., 1968.

National Tuberculosis and Respiratory Disease Association, *Chronic Obstructive Pulmonary Disease—A Manual for Physicians.* Portland, Oregon: Oregon Thoracic Society, 1966.

PARNELL, R. W., *Behavior and Physique*. London, England: Edward Arnold Publishers, Ltd., 1958.

SHELDON, W. H., *Atlas of Man*. New York: Harper & Row, Publishers, 1954.

SHELDON, W. H., S. S. STEVENS, and W. B. TUCKER, *Varieties of Human Physique*. New York: Harper & Row, Publishers, 1951.

VAN HUSS, WAYNE, et al., *Physical Activity in Modern Living* (2nd ed.). Englewood Cliffs, N.J.: Prentice-Hall, Inc., 1969.

chapter 3

CASADY, DONALD R., DONALD F. MAPES, and LOUIS E. ALLEY, *Handbook of Physical Fitness Activities*. New York: The Macmillan Company, Publishers, 1965.

DE VRIES, HERBERT A., *Physiology of Exercise*. Dubuque, Iowa: William C. Brown Company, 1966.

HEIN, FRED V., and ALLAN J. RYAN, "The Contributions of Physical Activity to Physical Health," *Research Quarterly*, 31 (1960), 263–85.

KEEN, MARTIN, *The Human Body*. New York: Grosset & Dunlap, Inc., 1961.

MAYER, JEAN, *Overweight*. Englewood Cliffs, N.J.: Prentice-Hall, Inc., 1968.

MONTOYE, H. J., "Summary of Research on the Relationship of Exercise to Heart Disease," *Journal of Sports Medicine and Physical Fitness*, 2 (1962), 35–43.

RATCLIFF, J. D., "How to Avoid Harmful Stress," *Reader's Digest*, July, 1970. Condensed from *Today's Health* published by the American Medical Association.

STAFF OF FITNESS FOR LIVING, *How to Keep Fit Throughout the Year*. Emmaus, Pennsylvania: Rodale Press, Inc., 1968.

——, *You're Never Too Old*. Emmaus, Pennsylvania: Rodale Press, Inc., 1967.

chapter 4

AHLMAN, K., and M. J. KARVONEN, "Weight Reduction by Sweating in Wrestlers and Its Effect on Physical Fitness," *Journal of Sports Medicine and Physical Fitness*, 1 (1962), 58–62.

ASPREY, G. M., et al., "Effect of Eating at Various Times on Subsequent Performance in the 440-Yard Dash and Half-Mile Run," *Research Quarterly*, 34 (1963), 267–70.

BERBER, R. A., "Mean Strength Resulting From Weight Training Programs Involving Six Different Methods," *Research Quarterly*, 33 (1962), 334.

BOWERMAN, BILL, "The Secrets of Speed," *Sports Illustrated*, 35, No. 5 (August 1971), 22–29.

CASADY, DONALD R., DONALD F. MAPES, and LOUIS E. ALLEY, *Handbook of*

Physical Fitness Activities. New York: The Macmillan Company, Publishers, 1965.

COOPER, KENNETH H., M.D., *Aerobics.* New York: M. Evans & Company, Inc., 1968.

————, *The New Aerobics.* New York: M. Evans & Company, Inc., 1970.

DE VRIES, HERBERT A., "Evaluation of Static Stretching Procedures for Improvement of Flexibility," *Research Quarterly,* 33 (1962), 222–29.

————, *Physiology of Exercise.* Dubuque, Iowa: William C. Brown Company, 1966.

KARPOVICH, PETER, and WAYNE E. SINNING, *Physiology of Muscular Activity* (7th ed.). Philadelphia: W. B. Saunders Company, 1971.

KARVONEN, M. J., "Effects of Vigorous Exercise on the Heart," in *Work and the Heart,* eds. F. F. Rosenbaum and E. L. Belknap. New York: Paul B. Hoeber, Inc., 1959.

KASCH, FRED W., and JACK BOYER, M.D. *Adult Fitness—Principles and Practices.* Greeley, Colorado: All-American Productions and Publications, 1968.

NADEL, J. A., and J. H. COMROE JR., "Acute Effects of Inhalation of Cigarette Smoke on Airway Conductance," *Journal of Applied Physiology,* 16 (1961), 713–16.

VAN HUSS, WAYNE, et al., *Physical Activity in Modern Living* (2nd ed.). Englewood Cliffs, N.J.: Prentice-Hall, Inc., 1969.

chapter 5

American Association for Health, Physical Education, and Recreation, *Weight Training in Sports and Physical Education.* Washington, D.C., 1962.

BOWERMAN, WILLIAM J., and W. E. HARRIS, M.D., *Jogging—A Physical Fitness Program for All Ages.* New York: Grosset and Dunlap, Inc., 1967.

CASADY, DONALD R., DONALD F. MAPES, and LOUIS E. ALLEY, *Handbook of Physical Fitness Activities.* New York: The Macmillan Company, Publishers, 1965.

COOPER, KENNETH H., *The New Aerobics.* New York: M. Evans & Company, Inc., 1970.

Exer-Genie: Revolutionary Exerciser (Instruction Manual). Fullerton, California: Exer-Genie, Inc., 1966.

FALLS, HAROLD B., E. L. WALLIS, and S. A. LOGAN, *Foundations of Conditioning.* New York: Academic Press, Inc., 1970.

HEALTHWAYS PHYSICAL FITNESS STAFF, *Health, Strength, and Fitness.* Los Angeles: Healthways, Inc., 1964. Basic and advanced weight training courses.

JOHNSON, HARRY J., M.D. with RALPH BASS, *Creative Walking for Physical Fitness.* Grosset and Dunlap, Inc., 1970.

KASCH, FRED W., and JACK BOYER, M.D. *Adult Fitness—Principles and Prac-*

tices. Greeley, Colorado: All-American Productions and Publications, Inc., 1968.

LEIGHTON, JACK R., *Progressive Weight Training.* New York: Ronald Press Company, 1961.

MITCHELL, CURTIS, "Tone Up The Swimming Pool Way," *Family Health,* July, 1970.

OLSON, EDWARD C., *Conditioning Fundamentals.* Columbus, Ohio: Charles E. Merrill Publishing Company, 1968.

P-F: A Physical Fitness Plan. Durham, North Carolina: Duke University Department of Health and Physical Education, 1967.

PRUDDEN, BONNIE, *Quick Rx for Fitness.* New York: Grosset and Dunlap, Inc., 1965. Exercises for women.

Royal Canadian Air Force Exercise Plans for Physical Fitness, rev. U.S. Edition. New York: Pocket Books, Inc., 1962.

SEATON, DON C., et al., *Physical Education Handbook* (5th ed.). Englewood Cliffs, N.J.: Prentice-Hall, Inc., 1969.

STEINHOUSE, ARTHUR, *How to Keep Fit and Like It* (2nd ed.), revd. and enlarged. Chicago: The Darnell Corporation, 1957.

chapter 6

ADAMS, W. C., et al., *Foundations of Physical Activity* (2nd ed.). Champaign, Illinois: Stipes Publishing Company, 1965.

BARROW, HAROLD M., MARJORIE CRISP, and JAMES W. LONG, *Physical Education Syllabus* (4th ed.). Minneapolis, Minnesota: Burgess Publishing Company, 1967.

HAMILTON, R. A., *Ready Reference for Adapted Exercise.* Auburn, New York: Reedco, 1967.

HICKMAN, CLEVELAND P., *Health for College Students* (3rd ed.). Englewood Cliffs, N.J.: Prentice-Hall, Inc., 1964.

KRAUS, HANS, M.D., and WILHEIM RAAB, M.D., *Hypokinetic Disease.* Springfield, Illinois: Charles C. Thomas, Publisher, 1961.

LOWMAN, CHARLES LE ROY, and CARL YOUNG, *Postural Fitness.* Philadelphia: Lea & Febiger, 1960.

chapter 7

ADAMS, W. C., et al., *Foundations of Physical Activity* (2nd ed.). Champaign, Illinois: Stipes Publishing Company, 1965.

American Medical Association, *Vitamin Supplements and Their Correct Use.* Department of Foods and Nutrition, 1964.

BOGERT, L. JEAN, GEORGE M. BRIGGS, and DORIS HOWES CALLOWAY, *Nutrition and Physical Fitness* (8th ed.). Philadelphia: W. B. Saunders Company, 1966.

CAMERON, A. G., *Food and Its Function* (2nd ed.). London: Edward Arnold Publisher, 1968.

HEWITT, DONALD W., M.D., *Reduce and Be Happy.* Mountain View, California: Pacific Press Publishing Association, 1955.

JOHNS HOPKINS HOSPITAL, *Manual of Applied Nutrition* (5th ed.). Baltimore: The Johns Hopkins Press, 1966.

MAYER, JEAN, *Overweight.* Englewood Cliffs, N.J.: Prentice-Hall, Inc., 1968.

THURSTON, EMORY W., "Are Vitamins Necessary?" *Let's Live,* 37 (May 1969), 36.

U.S. Department of Agriculture, *Nutritive Value of Foods.* Revised Home and Garden Bulletin No. 72. Washington, D.C.: U.S. Government Printing Office, 1970.

Vanderbilt University Hospital Dietary Staff, *Diet Manual* (2nd ed.). Nashville: Vanderbilt University Press, 1969.

VAN HUSS, WAYNE, et al., *Physical Activity in Modern Living* (2nd ed.). Englewood Cliffs, N.J.: Prentice-Hall, Inc., 1969.

chapter 8

DE VRIES, HERBERT A., *Physiology of Exercise.* Dubuque, Iowa: William C. Brown Company, 1966. See especially chapter on "Prophylactic and Therapeutic Effects of Exercise.

JACOBSON, E., *Progressive Relaxation.* Chicago: University of Chicago Press, 1956.

JOHNSON, NORA, "Keeping Perky in a Pressure Cooker," *Cosmopolitan,* July, 1970, pp. 147–49.

KNIGHT, LEAVITT A., "How to Relax Without Pills," *Readers Digest,* February, 1971. Condensed from *The American Legion Magazine,* October, 1970.

LUCE, GAY GAER, and JULIUS SEGAL, *Sleep.* New York: Coward-McCann, Inc., 1966.

MENNINGER, W. C., "Recreation and Mental Health," *Recreation,* 42 (November 1968), 340–46.

RATCLIFF, J. D., "How to Avoid Harmful Stress," *Readers Digest,* July, 1970. Condensed from *Today's Health,* July, 1970.

SELYE, HANS, *The Stress of Life.* New York: McGraw-Hill Book Company, 1956.

———, "Unmasking the Faces of Stress," *Medical Tribune,* October, 1961.

STEARN, JESS, *Yoga, Youth, and Reincarnation.* New York: M. Evans & Company, Inc., 1968.

STEINROHN, PETER, J., M.D., *How to Get a Good Night's Sleep.* Chicago: Henry Regnery Co., 1968.

Van Huss, Wayne, et al., *Physical Activity in Modern Living* (2nd ed.). Englewood Cliffs, N.J.: Prentice-Hall, Inc., 1969.

chapter 9

Knapp, Clyde, and E. Patricia Hagman, *Teaching Methods for Physical Education.* New York: McGraw-Hill Book Company, 1953.

Lawther, John D., *The Learning of Physical Skills.* Englewood Cliffs, N.J.: Prentice-Hall, Inc., 1968.

Murcell, James L., *Successful Teaching* (2nd ed.). New York: McGraw-Hill Book Company, 1954.

Penman, Kenneth A., *Physical Education for College Students.* St. Louis: The C. V. Mosby Co., 1967.

Van Huss, Wayne, et al., *Physical Activity in Modern Living* (2nd ed.). Englewood Cliffs, N.J.: Prentice-Hall, Inc., 1969.

appendix b

Sinacore, John S., *Health—A Quality of Life.* New York: The Macmillan Company, Publishers, 1968.

The Sentinel, San Diego, California, "How I Quit Smoking," a 1971 series of letters from ex-smokers with comments by medical personnel of the San Diego Council on Smoking and Health, including the San Diego County Medical Society, The San Diego County Psychological Association, and the University of California, San Diego, School of Medicine.

index